The Handbook of Environmental Chemistry

Editors-in-Chief: O. Hutzinger · D. Barceló · A. Kostianoy

Volume 2 Reactions and Processes
Part P

The Handbook of Environmental Chemistry

Recently Published and Forthcoming Volumes

Polymers: Chances and Risks
Volume Editors: P. Eyerer, M. Weller and C. Hübner
2010

Alpine Waters
Volume Editor: U. Bundi
Vol. 6, 2010

The Aral Sea Environment
Volume Editors: A. G. Kostianoy and A. N. Kosarev
2010

Transformation Products of Synthetic Chemicals in the Environment
Volume Editor: A. B. A. Boxall
Vol. 2/P, 2009

Contaminated Sediments
Volume Editors: T. A. Kassim and D. Barceló
Vol. 5/T, 2009

Biosensors for the Environmental Monitoring of Aquatic Systems
Bioanalytical and Chemical Methods for Endocrine Disruptors
Volume Editors: D. Barceló and P.-D. Hansen
Vol. 5/J, 2009

Environmental Consequences of War and Aftermath
Volume Editors: T.A. Kassim and D. Barceló
Vol. 3/U, 2009

The Black Sea Environment
Volume Editors: A. Kostianoy and A. Kosarev
Vol. 5/Q, 2008

Emerging Contaminants from Industrial and Municipal Waste
Removal Technologies
Volume Editors: D. Barceló and M. Petrovic
Vol. 5/S/2, 2008

Emerging Contaminants from Industrial and Municipal Waste
Occurrence, Analysis and Effects
Volume Editors: D. Barceló and M. Petrovic
Vol. 5/S/1, 2008

Fuel Oxygenates
Volume Editor: D. Barceló
Vol. 5/R, 2007

The Rhine
Volume Editor: T. P. Knepper
Vol. 5/L, 2006

Persistent Organic Pollutants in the Great Lakes
Volume Editor: R. A. Hites
Vol. 5/N, 2006

Antifouling Paint Biocides
Volume Editor: I. Konstantinou
Vol. 5/O, 2006

Estuaries
Volume Editor: P. J. Wangersky
Vol. 5/H, 2006

The Caspian Sea Environment
Volume Editors: A. Kostianoy and A. Kosarev
Vol. 5/P, 2005

Marine Organic Matter: Biomarkers, Isotopes and DNA
Volume Editor: J. K. Volkman
Vol. 2/N, 2005

Transformation Products of Synthetic Chemicals in the Environment

Volume Editor: Alistair B. A. Boxall

With contributions by

C. D. Adams · D. Barceló · W. A. Battaglin · R. Baumgartner
A. B. A. Boxall · J. Coats · K. E. Conn · L. B. M. Ellis
B. I. Escher · K. Fenner · E. T. Furlong · S. T. Glassmeyer
K. Henderson · P. H. Howard · D. Hu · S. J. Kalkhoff · D. W. Kolpin
J. Lienert · M. T. Meyer · S. Pérez · M. Petrovic · U. Schenker
M. Scheringer · D. J. Schnoebelen · C. J. Sinclair · L. P. Wackett

Springer

Environmental chemistry is a rather young and interdisciplinary field of science. Its aim is a complete description of the environment and of transformations occurring on a local or global scale. Environmental chemistry also gives an account of the impact of man's activities on the natural environment by describing observed changes.
The Handbook of Environmental Chemistry provides the compilation of today's knowledge. Contributions are written by leading experts with practical experience in their fields. The Handbook will grow with the increase in our scientific understanding and should provide a valuable source not only for scientists, but also for environmental managers and decision-makers.
The Handbook of Environmental Chemistry is published in a series of five volumes:

Volume 1: The Natural Environment and the Biogeochemical Cycles
Volume 2: Reactions and Processes
Volume 3: Anthropogenic Compounds
Volume 4: Air Pollution
Volume 5: Water Pollution

The series Volume 1 The Natural Environment and the Biogeochemical Cycles describes the natural environment and gives an account of the global cycles for elements and classes of natural compounds.
The series Volume 2 Reactions and Processes is an account of physical transport, and chemical and biological transformations of chemicals in the environment.
The series Volume 3 Anthropogenic Compounds describes synthetic compounds, and compound classes as well as elements and naturally occurring chemical entities which are mobilized by man's activities.
The series Volume 4 Air Pollution and Volume 5 Water Pollution deal with the description of civilization's effects on the atmosphere and hydrosphere.
Within the individual series articles do not appear in a predetermined sequence. Instead, we invite contributors as our knowledge matures enough to warrant a handbook article.
Suggestions for new topics from the scientific community to members of the Advisory Board or to the Publisher are very welcome.

The Handbook of Environmental Chemistry, Subseries 2 ISSN 1433-6839
ISBN 978-3-642-26041-4 e-ISBN 978-3-540-88273-2
DOI 10.1007/978-3-540-88273-2
Springer Dordrecht Heidelberg London New York

Cover design: WMXDesign GmbH, Heidelberg
Typesetting and Production: le-tex publishing services GmbH, Leipzig

Printed on acid-free paper

Springer is part of Springer Science+Business Media (www.springer.com)

Editors-in-Chief

Prof. em. Dr. Otto Hutzinger

Universität Bayreuth
c/o Bad Ischl Office
Grenzweg 22
5351 Aigen-Vogelhub, Austria
hutzinger-univ-bayreuth@aon.at

Prof. Andrey Kostianoy

P.P. Shirshov Institute of Oceanology
Russian Academy of Sciences
36, Nakhimovsky Pr.
117997 Moscow, Russia
kostianoy@mail.mipt.ru

Prof. Dr. Damià Barceló

Department of Environmental Chemistry
IDAEA-CSIC, C/Jordi Girona 18–26,
08034 Barcelona, Spain, and Catalan
Institute for Water Research (ICRA),
Parc Científic i Tecnològic de la
Universitat de Girona,
Edifici Jaume Casademont, 15
E-17003 Girona, Spain
dbcqam@iiqab.csic.es

Volume Editor

Dr. Alistair B.A. Boxall

Environment Department
University of York
Heslington, York, YO10 5DD
United Kingdom
alistair.boxall@fera.gsi.gov.uk

Advisory Board

Prof. Dr. J. P. Giesy
Department of Zoology
Michigan State University
East Lansing, MI 48824-1115, USA
Jgiesy@aol.com

Prof. Dr. R. A. Hites
Indiana University
School of Public
and Environmental Affairs
Bloomington, IN 47405, USA
hitesr@indiana.edu

Prof. Dr. M. A. K. Khalil
Department of Physics
Portland State University
Science Building II, Room 410
P.O. Box 751
Portland, OR 97207-0751, USA
aslam@global.phy.pdx.edu

Prof. Dr. D. Mackay
Department of Chemical Engineering
and Applied Chemistry
University of Toronto
Toronto, ON, M5S 1A4, Canada

Prof. Dr. A. H. Neilson
Swedish Environmental Research Institute
P.O. Box 21060
10031 Stockholm, Sweden
ahsdair@ivl.se

Prof. Dr. J. Paasivirta
Department of Chemistry
University of Jyväskylä
Survontie 9
P.O. Box 35
40351 Jyväskylä, Finland

Prof. Dr. Dr. H. Parlar
Institut für Lebensmitteltechnologie
und Analytische Chemie
Technische Universität München
85350 Freising-Weihenstephan, Germany

Prof. Dr. S. H. Safe
Department of Veterinary
Physiology and Pharmacology
College of Veterinary Medicine
Texas A & M University
College Station, TX 77843-4466, USA
ssafe@cvm.tamu.edu

Prof. P. J. Wangersky
University of Victoria
Centre for Earth and Ocean Research
P.O. Box 1700
Victoria, BC, V8W 3P6, Canada
wangers@telus. net

The Handbook of Environmental Chemistry Also Available Electronically

For all customers who have a standing order to The Handbook of Environmental Chemistry, we offer the electronic version via SpringerLink free of charge. Please contact your librarian who can receive a password or free access to the full articles by registering at:

springerlink.com

If you do not have a subscription, you can still view the tables of contents of the volumes and the abstract of each article by going to the SpringerLink Homepage, clicking on "Browse by Online Libraries", then "Chemical Sciences", and finally choose The Handbook of Environmental Chemistry.

You will find information about the

- Editorial Board
- Aims and Scope
- Instructions for Authors
- Sample Contribution

at springer.com using the search function.

Color figures are published in full color within the electronic version on SpringerLink.

Preface

Following release to the environment, synthetic chemicals may be degraded by biotic and abiotic processes. The degradation of the chemical can follow a plethora of pathways and a range of other substances can be formed via these different pathways (e.g. [1]). A number of terms have been used for these substances including metabolites, degradates and transformation products – in this book we use the term transformation products. While we often know a lot about the environmental properties and effects of the parent synthetic chemical, we know much less about the transformation products.

Transformation products can behave very differently from the parent compound (e.g. [2]). For example, selected transformation products are much more persistent than their associated parent compound in soils, waters and sediments and some may be transported around the local, regional and global environments to a different extent than the parent compound. Transformation products can also have very different toxicities than the parent compound (e.g. [3]) and in some cases transformation products can be orders of magnitude more toxic than their parent compound; although this situation is rare. The environmental risks of transformation products can therefore be very different than the risks of the parent compound.

The potential environmental impacts of transformation products are recognised by many regulatory assessment schemes. For example, in the EU, pesticide producers are not only required to assess the fate and effects of the parent pesticide but are also required to assess the potential adverse effects of major metabolites and minor metabolites that are deemed to be of concern [4]. Similar requirements also exist for new human and veterinary pharmaceuticals and biocides (e.g. [5]). However, for many older substances and many other substance classes (e.g. industrial chemicals), data on the environmental risks of transformation products can be limited or non-existent.

The assessment of the environmental risks of transformation products can however be challenging. Perhaps the biggest challenge is that there are a vast number of synthetic chemicals in use today which can each degrade into a number of transformation products; we don't have the resources to test the fate and environmental effects of the parent compounds let alone the transformation products. The identification and characterisation of transformation products arising from a particular parent substance in a particular system can also be extremely difficult due to problems of extraction, detection at environmentally relevant levels, and quantification in the absence of standards; although the arrival of new analytical methodologies (e.g. time-of-flight mass spectrometry) and the availability of expert systems for predicting transformation pathways is now making this task less daunting. The modelling of transformation product exposure and effects can also be challenging as we are faced with a dynamic system involving a complex mixture of substances where parent compounds are being degraded to transformation products which are then degraded to other transformation products. Finally, while treatment methodologies that are used to control human and environmental exposure are able to remove transformation products, they can also act as a mechanism of transformation product formation and selected treatment processes (e.g. advanced oxidation processes for drinking water treatment) may even produce transformation products more hazardous than the substance that has been treated.

While, there are a number of scientific challenges and large knowledge gaps, a significant amount of information is available on the routes of formation, detection, exposure, effects and modelling approaches for transformation products of some classes of substances. If we can bring this information together, we should be able to assess transformation products in a much more pragmatic way. This will allow resources to be focused on transformation products of most concern while maintaining the health of the natural environment.

Therefore in this book, we have brought together contributions from leading experts in this field to provide an overview of the current knowledge on the formation, detection, occurrence, effects and treatability of transformation products in the environment. Many of the chapters introduce methods for assessing the different components required to determine the risks of transformation products to natural systems. In the chapter *Mechanisms of degradation of synthetic chemicals*, Wackett et al. (this volume) discuss the mechanisms by which transformation products are formed and describe how this information can be used to predict the structures of transformation products. Howard discusses a wider range of methods for predicting degradation rates and degradation pathways in the chapter *Predicting the persistence of organic compounds*. The chapter *Analysing transformation products of synthetic chemicals* by Perez et al. describes the challenges for analysing transformation products and discusses the application of some of the new analytical methods for identification and quantification of transformation products in environmental systems. *In Occurrence of Transformation Products in the Environment*, Kolpin describes the results of a series of monitoring studies into the occurrence of selected transformation in US water bodies. Hu et al. (*Fate of Transformation Products of Synthetic Chemicals*) discuss experimental data on the persistence and mobility of transformation products in environmental systems and in the chapter *Modeling environmental exposure to transformation products of organic chemicals*, Fenner et al. describe modelling approaches for assessing exposure levels for transformation products in a range of environmental systems. The chapters

Ecotoxicity of Transformation Products (Sinclair and Boxall) and *Predicting the Ecotoxicological Effects of Transformation Products* (Escher et al.) describe the ecotoxicological effects of transformation products and discuss approaches that could be employed for estimating ecotoxicity based on transformation product structure and information on the associated parent chemicals. Finally, in *Treatment of Transformation Products*, Adams et al. discuss how transformation products can be removed in treatment processes but also discuss how treatment processes can act as routes of transformation product formation.

It is clear from each of the chapters that while we are now well placed to better assess transformation product risk, there is still much that needs to be done. Areas where we need further development include:

- Expert systems for predicting the nature of transformation products – Work should focus on the development of methods to identify the most probable transformation pathway in a particular environmental system. The approaches need to be evaluated against high-quality experimental data on degradation pathways in different media. New expert systems need to be developed for systems where they are not yet available, e.g. drinking water treatment processes.
- Analytical methods – We need to develop high-quality methods that are able to extract and identify all transformation products of potential concern in a range of environmental systems. We should explore how we can quantify (or semi-quantify) transformation product concentrations in the absence of standards.
- Monitoring studies for transformation products – A number of monitoring studies have explored the occurrence of transformation products in the environment. These studies have tended to focus on transformation products arising from the use of only a few pesticide active ingredients. It would be useful to prioritise transformation products in terms of their potential risk to a particular system (e.g. using approaches similar to that described by Sinclair et al. [6]) and extend these monitoring studies to a much wider range of substances. Where possible, monitoring studies should not just look at occurrence but should also aim to understand the underlying mechanisms determining the transport of transformation products around the environment.
- Exposure models – Models are available for estimating exposure of transformation products at a range of scales. These models need evaluation and may need further development as our knowledge expands.
- Ecotoxicological effects – Most experimental data is on the acute toxicity of transformation products to aquatic organisms so it would be valuable to generate an understanding of the potential chronic effects as well as an understanding of the impacts on terrestrial organisms. Predictive approaches for estimating the ecotoxicity of transformation products show some promise, however these need further development and validation. It

is also important to recognise that a transformation product will not occur in the environment on its own but will co-occur with its parent compound, other parent compounds and other transformation products, the further development of approaches for assessing the risk of mixtures is therefore critical. As the system is a dynamic system (i.e. concentrations of parent compounds and transformation products will be changing at different rates), in the future mixture assessment models that can deal with changing exposure concentrations may be required.
- Human health implications of transformation products – Most work to date has focused on the assessment and prediction of the ecotoxicity of transformation products. We need to begin to assess the potential human health implications of the presence of transformation products in the environment and develop approaches for identifying transformation products of most concern to human health. Expert systems for predicting mammalian toxicity endpoint may play a role here.

To address these issues will require input from a wide range of disciplines including ecotoxicologists, exposure modellers, analytical chemists, toxicologists, treatment scientists and biochemists. Hopefully this book will encourage researchers, students and regulators from these different fields to begin, or continue, to work to develop approaches and knowledge so that in the future we have a much better understanding of the risks of transformation products and of how to control these risks.

Heslington, York, June 2009 Alistair Boxall

References

1. Roberts, T.; Hutson, D. *Metabolic Pathways of Agrochemicals, Part Two: Insecticides and Fungicides*; The Royal Society of Chemistry: Cambridge, 1999.
2. Boxall ABA, Sinclair CJ, Fenner K, Kolpin D, Maund SJ (2004) Environ. Sci. Technol. 38:368A
3. Sinclair CJ, Boxall ABA (2003) Environ. Sci. Technol. 37:4617
4. European Commission, Guidance Document on Aquatic Ecotoxicology in the Context of the Directive 91/414/EEC, Sanco/3268/2001 rev.4 (final), Brussels, 2002.
5. VICH, Environmental Impact Assessments for Veterinary Medicinal Products - Phase II, VICH GL38, International Cooperation on Harmonization of Technical Requirements for Registration of Veterinary Products, 2004.
6. Sinclair CJ, Boxall ABA, Parsons SA, Thomas MR (2006) Environ Sci Technol 40: 7283

Contents

Part III: Effects of Transformation Products

Part I
Formation, Detection and Occurrence of Transformation Products

Hdb Env Chem Vol. 2, Part P (2009): 3–16
DOI 10.1007/698_2_014

Published online: 14 March 2008

Mechanisms of Degradation of Synthetic Chemicals

Lawrence P. Wackett[1] (✉) · Lynda B. M. Ellis[2]

[1]Department of Biochemistry, Molecular Biology, and Biophysics and BioTechnology Institute, University of Minnesota, 1479 Gortner Avenue, St. Paul, MN 55108, USA
wacke003@umn.edu

[2]Department of Laboratory Medicine and Pathology, Minneapolis, MN 55455, USA

Abstract The fate of chemicals in the environment is largely dependent upon microbial biodegradation, or a lack thereof. Biodegradation derives from the extremely broad types of metabolic reactions catalyzed by microbes. Diverse microbial metabolism is represented in the University of Minnesota Biocatalysis/Biodegradation Database (UM-BBD), which is freely available on the Web. The UM-BBD encompasses metabolism of 60 organic functional groups. On average, there are four reaction types for each functional group. Each of these reaction types underlies a metabolic rule. Metabolic rules have formed the basis of a computational system used to predict the biodegradative pathways of chemicals. Many pathways may be predicted. To deal with pathway combinatorial explosion, a rule-prioritization system has been implemented. Additional tools are under development to better understand the underlying characteristics of biodegradative metabolism with the hope of improving biodegradation prediction.

Keywords Biodegradation · Database · Metabolism · Microbes · Pathways · Prediction

Abbreviations

ACA	L-aztidine-2-carboxylic acid
ATP	Adenosine triphosphate
BESS	Biodegradation evaluation and simulation system
HAD	2-Haloacid dehalogenase
PCBs	Polychlorinated biphenyls
PCE	Perchloroethylene
PPS	Pathway prediction system
SMILES	Simplified molecular input line entry system
UM-BBD	University of Minnesota Biocatalysis/Biodegradation Database

1 Introduction

There are approximately 87 000 chemical substances in the United States EPA registration of commercial compounds [1]. This includes relatively simple molecules like methanol, and much more complex molecules, for example those found in personal care products and pharmaceuticals. While the latter are often present in the environment in rather low concentrations, their strong biological activity may give cause for concern.

In general, the fate of commercial chemicals in the environment is predicated on the ability of microorganisms to metabolize them. However, only a small fraction of these 87 000 chemical substances have documented information in peer-reviewed scientific journals on their biodegradation by microbes. This gap between chemical and microbial metabolic information will increase over time since chemists make new substances for deployment by industry at a faster rate than studies on biodegradation of new substances are conducted. This necessitates a better understanding of the underlying principles of microbial biodegradative metabolism. These principles can be used to predict how new substances may be degraded. Regulators are increasingly requiring degradation rate and route studies as part of the environmental risk assessment of pesticides, pharmaceuticals, biocides, and veterinary medicines (see Chap. 1) and such knowledge will also be invaluable in guiding the performance of these studies.

2 Significance of Microbial Biodegradation

Microbial metabolism is highly diverse with respect to the mechanisms and substrate specificity displayed by the enzymes that mediate the individual reactions. This statement is based on direct elucidation of metabolism in the laboratory, and indirectly by the disappearance of chemicals in the environment following a suitable biological "acclimation" period. There are ample cases where compounds, once thought to be non-biodegradable, were later found to be degraded. One notable example is polychlorinated biphenyl (PCB) congeners containing 8, 9, and 10 chlorine substituents. These were initially considered to persist in the environment. However, later these highly chlorinated congeners were observed to disappear with the concomitant appearance of congeners containing hydrogen atoms in place of one or more displaced chlorine atoms [2]. Subsequent work clearly established the process to be microbially mediated [3]. Other workers obtained pure microbial cultures that catalyze reductive dechlorination of chlorinated aromatic rings distinct from PCBs. In some cases, chlorinated organic compounds have been shown to act as final electron acceptors in microbial metabolism. The net effect of this reductive metabolism is to contribute to the degradation of the chlorinated compounds. Reductive dehalogenases have been purified and studied in vitro [4, 5].

Microbial metabolism, or the lack thereof, determines whether chemicals persist or leave the environment quickly. When organic compounds are completely metabolized to carbon dioxide, then any detrimental environmental effect the chemical might manifest is alleviated. In some cases, for example with the PCBs described above, incomplete metabolism occurs. For the majority of substances, the resulting degradation products will be less toxic to humans and the environment than the parent compound (see chapters by Sinclair and Boxall and Escher et al.). For example, it is beneficial for PCBs to undergo even partial reductive dechlorination; less chlorinated PCB congeners are generally less toxic and persistent than their more heavily chlorinated counterparts.

In another example, however, the chlorinated solvent perchloroethylene (PCE) has been observed to undergo microbially mediated, partial dechlorination in the environment to generate dichloroethenes and vinyl chloride [6]. The latter is a potent human carcinogen [7]. The starting compound PCE is considered to be much less harmful. In this case, microbial metabolism can generate a metabolite that is significantly worse than PCE, a compound that is used in dry-cleaning operations around the globe. Other examples where a transformation process results in an increase in toxicity to humans and the environment are given in later chapters (see chapters by Sinclair and Boxall and Escher et al.).

For all of the above reasons, it is important to predict the fate of chemicals in the environment by predicting the course of microbial metabolism, and it is clearly necessary to know more than whether the compound is metabolized but also which metabolic route or routes the degradation process follows.

3 University of Minnesota Biocatalysis/Biodegradation Database

For more than 12 years, the University of Minnesota Biocatalysis/Biodegradation Database (UM-BBD) has grown, so that it now includes a good deal of the breadth of microbial metabolic reactions, particularly those important in determining the fate of chemicals in the environment [8]. The UM-BBD contains information on environmental chemical compounds (e.g., pesticides), microorganisms (primarily bacteria), enzyme reactions, and metabolic pathways. The information in the database can be searched in numerous ways and retrieved in different forms. There are extensive links to further information on enzymes, gene sequences, and other relevant information. Some of the reactions depicted on the UM-BBD are significant for their use for synthesis of specialty and commodity chemicals. For example, nitrile hydratase, important in the biodegradation of the pesticide bromoxynil, is also used industrially to synthesize acrylamide from acrylonitrile [9].

A compilation of biocatalytic and biodegradative reactions has value to users in a number of domains. The information is useful to regulatory agencies that consider the fate of chemicals in the environment as part of their activities. Knowledge of biodegradation reactions is now considered by many chemical manufacturers when they design compounds for environmental impact; for example, with pesticides. The goal is to produce molecules with a specific beneficial activity that do not degrade too quickly, but also do not persist indefinitely. It is also important to understand if a non-toxic chemical may be degraded to a chemical intermediate that is toxic and accumulates. Knowledge of biodegradation reactions is important to researchers who seek to uncover new biodegradative metabolism, or to use the information for related activities like bioremediation. Information on degradation pathways is also now required by many regulatory risk-assessment schemes. There is also a value in compiling information in any field for the purpose of defining what we don't know. We have done this by analyzing UM-BBD reactions in the context of delineating the chemical functional groups that exist in nature, and determining which of those are currently covered by existing knowledge that is captured within the UM-BBD.

4
Chemical Functional Groups

The reactions in the UM-BBD can be organized in many different ways. One of the most useful, in the context of understanding metabolism globally, is to organize the reactions based on the chemical functional group undergoing transformation [10]. Thus, in any given reaction, one functional group is undergoing a metabolic transformation to another functional group. For example, an alcohol may be oxidized to an aldehyde and the aldehyde can, in turn, be oxidized to a carboxylic acid (Fig. 1). So the terminal functional group of one biotransformation becomes the starting material for the next. The string of functional group transformations form biodegradative metabolic pathways.

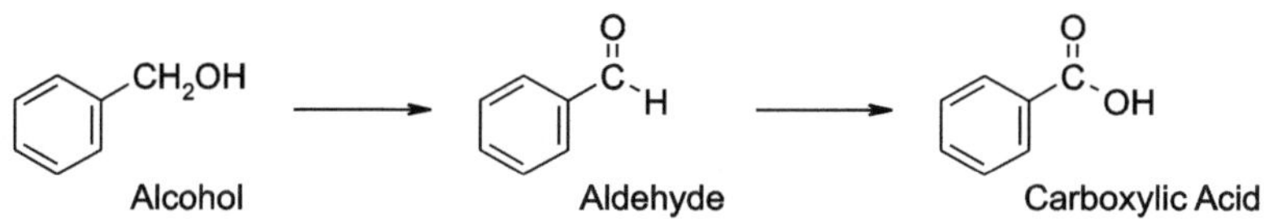

Fig. 1 Functional group transformations of the type giving rise to rules

5
Microbial Metabolic Breadth

Microbes carry out all of the functional group transformations that occur in higher organisms and many additional ones. Certain functional groups are only known to be transformed by microbial enzymes; and metabolism either does not exist, or has not yet been discovered, in higher organisms. For example, organomercurials are metabolized by an organomercurial lyase, MerB, found only in prokaryotes [11]. To our knowledge, only microbes productively metabolize organosilicon [12], organoboron [13] and organotin compounds [14]. Microbial metabolism is thought to represent the greatest breadth of metabolism on earth [15].

How do we estimate the breadth of microbial metabolism? We start with the well-documented reactions for 60 functional groups found in metabolism contained within the UM-BBD. Next, we ask what other functional groups might exist in nature, for which knowledge of their chemical transformation is currently lacking. If a functional group is found within a natural product, for example an antibiotic, then biochemistry clearly exists to biosynthesize that functional group. Furthermore, since natural products are recycled metabolically, there must also exist catabolic reactions for the biodegradation of the particular functional group. The set of functional groups that exists in natural products, but for which metabolic transformations thereof are lacking, defines some of what we do not know about biodegradation. We believe that an important research enterprize in any field is to define what is not known, which constitutes a roadmap for future discovery.

The natural product literature contains approximately 100 functional groups [15]. Most biochemistry textbooks deal with approximately 30; the UM-BBD covers 60. These constitute somewhat less than 30% or 60%, respectively, of the functional groups known to exist in nature. These percentages are almost surely overestimates, as there are likely some functional groups that have not yet been reported in the literature for natural products. Since microbes recycle most organic matter on the planet, there is a great likelihood that these yet-to-be-investigated functional groups are metabolized by microbes.

Our appreciation of the extent of our ignorance of microbial metabolism is important in the context of the widespread sequencing and annotation of microbial genomes. The complete genome sequences for thousands of prokaryotes have been, or are currently being, determined [16]. This is an exciting development that is considered to be on the forefront of biological and environmental science. The genome sequence itself is not the objective; rather it is the ability to translate the DNA sequence into gene sequences, and the gene sequences into protein sequences. The protein sequences are compared to protein sequences in databases and the putative functions of the proteins are assigned based on computer-based methods.

However, if we acknowledge our ignorance of a significant set of metabolic reactions that are carried out by microbes, it follows that many genes will not have functions ascribed to them properly. Consistent with this is the finding that 20 to 50% of genes identified by gene-finding programs are annotated as "unknown function" or "hypothetical" [17]. Additionally, it is widely acknowledged that there is a significant amount of misannotation, the attribution of incorrect function, with computationally annotated proteins [18]. This arises in part because proteins with related sequences catalyze different reactions, but precise function can often not be predicted by sequence information alone.

6
New Mechanisms in Biodegradation

In this genomic context, there is a strong need to continue to delineate new biochemical reactions. Some of these will be uncovered by identifying new biodegradation reactions. In the last several years, there have been new biochemical reactions determined in the biodegradation of novel chemical substances. A number of these new reactions have been added to the UM-BBD and expand the coverage of biochemical functional group metabolism. In some cases, gene and protein sequences for the enzymes have been determined and this also contributes to the annotation-function problem. Some specific examples that illustrate this point are discussed below.

6.1
Nitroaromatic Compounds

For example, new biochemical knowledge contributed to identifying a class of hypothetical proteins identified in wide-scale genome sequencing projects. The new biochemical knowledge derived from studies of Spain and coworkers on the microbial biodegradation of nitrobenzene compounds (Fig. 2). Nitrobenzene was shown to undergo reduction to nitrosobenzene and then further reduction to a hydroxylamine [19]. Hydroxylaminobenzene was shown to undergo enzyme-catalyzed isomerization to *o*-aminophenol, a reaction known in organic chemistry as the Bamberger-rearrangement [20]. The enzyme catalyzing this novel reaction was denoted as hydroxylaminobenzene

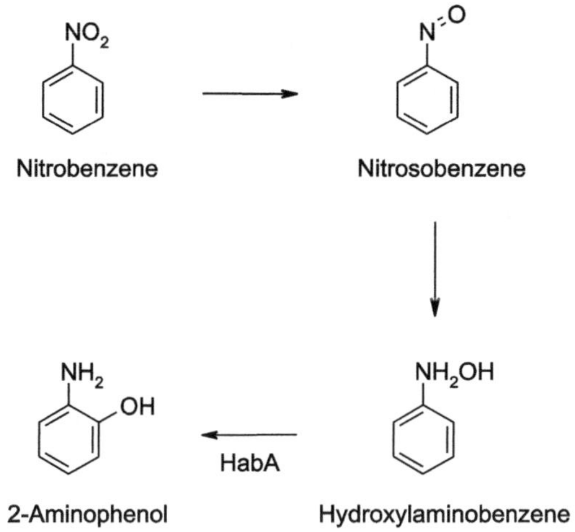

Fig. 2 Metabolic pathway for nitrobenzene with hydroxylaminobenzene mutase (HabA)

mutase [21]. The sequence of the *habA* gene from *Pseudomonas pseudoalcaligenes* JS45 was determined and the translated protein sequence was queried against GenBank using the BLAST algorithm [22]. The search identified dozens of homologous sequences that had been previously obtained via genome sequencing, but the genes had not been annotated because no biochemical function could be attributed to the sequence prior to the studies of Davis et al. [22]. One of the homologs of *habA* was a gene identified during the genome sequencing of *Mycobacterium tuberculosis*, the causative agent of tuberculosis. It is intriguing that this HabA homolog may have medical importance. Nitroaromatic compounds are an important class of anti-tubercular pro-drugs that are activated within *M. tuberculosis* via nitro group reduction to reactive hydroxylamino intermediates [23–25]. Isomerization of the hydroxylamine might constitute a resistance mechanism for *M. tuberculosis* strains, thus rendering HabA as a potential critical target for new compound drugs combating tuberculosis. In this example, the biochemical elucidation of HabA function may be important for genomics and medicine, as well as for biodegradation.

6.2 Azetidine Ring Compounds

In another example, biodegradation of azetidine ring structures was investigated using the toxic, plant, natural product L-azetidine-2-carboxylic acid (ACA). ACA is made in large quantities by certain plants, such as Lily of the Valley, to ward off pathogens [26]. Despite this, some bacteria metabolize ACA productively (Fig. 3). *Pseudomonas* strain A2C was isolated from soil beneath Lily of the Valley plants and was able to grow on L-azetidine-

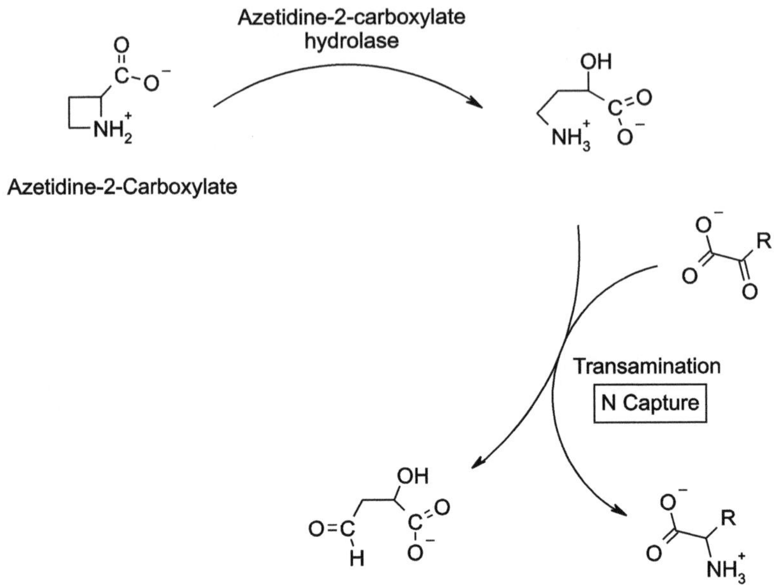

Fig. 3 Bacterial metabolism of azetidine-2-carboxylate initiated by ring-opening hydrolase enzyme

2-carboxylic acid as the sole nitrogen source. This suggests that the *Pseudomonas* strain A2C metabolizes ACA via a pathway cleaving the azetidine ring to produce nitrogen, and assimilates the nitrogen. ACA is similar to the amino acid L-proline but has one less methylene carbon. This mimicry of L-proline is the basis for the toxicity of ACA. Susceptible organisms take up ACA and incorporate it into proteins [27], but proteins have evolved for the specific bond angles of proline; many are inactive with ACA incorporated in place of proline.

ACA is biodegraded differently than proline, which is biodegraded by an initial oxidation reaction to generate an imine and the cyclic imine is then opened hydrolytically [28]. In *Pseudomonas* sp. A2C, the four-membered azetidine ring of ACA undergoes direct hydrolytic opening to produce 2-hydroxy-4-aminobutyrate (Fig. 3), which is non-toxic. 2-Hydroxy-4-aminobutyrate has been reported to undergo transamination to capture the amino group into cellular metabolism [29]. Thus, ACA metabolism both detoxifies a plant toxin and provides for cellular nitrogen capture at the same time. We have cloned and sequenced the ACA metabolism operon. The enzyme encoding ACA hydrolysis was identified. It is a member of the 2-haloacid dehalogenase (HAD) superfamily (C. Gross, unpubl. data), a very large protein superfamily. Many HAD genes have been identified by computer-based methods during microbial genome annotation. A recent study showed that many such genes, annotated as dehalogenases, do not encode proteins having dehalogenase activity. Some members of the HAD superfamily are known to be involved in the biodegradation of phosphonic acids [30]. Some members of this family, azetidine-2-carboxylate hydrolase enzymes, are now known to be involved in azetidine ring biodegradation.

6.3 Thioamide Compounds

The biodegradation of amides is well known but comparatively little work has been done on the biodegradation of thioamides. Commercial thioamides include the herbicide chlorthiamid, 2,6-dichlorothiobenzamide, and certain anti-tubercular drugs. One recent study showed that thioamides served as the sole nitrogen source supporting the growth of microbial enrichment cultures [31], but the mechanism underlying thioamide metabolism was not established in that study. More recently, thioacetamide was used as the sole nitrogen source to isolate a *Ralstonia picketti* strain from soil [32]. *R. picketti* strain TA metabolized both thiacetamide and thiobenzamide. With thiobenzamide, which did not support growth, benzamide and benzonitrile were observed to accumulate and thiobenzamide *S*-oxide was established to be an intermediate on the pathway. In general, thioamides are oxidized to thioamide *S*-oxide and a proposed second oxygenation step produces an unstable *S*-dioxo intermediate (Fig. 4). The *S*-dioxo intermediate is thought to undergo a spontaneous elimination reaction to generate either a nitrile or an amide, or a mixture of both. This was an unexpected set of reactions and illustrates the need to continue to discover novel mechanisms of microbial biodegradation. Pathways for the biodegradation of nitroaromatic and thioamide compounds are now contained within the UM-BBD. As new mechanisms of biodegradation are discovered, they will be added to the database.

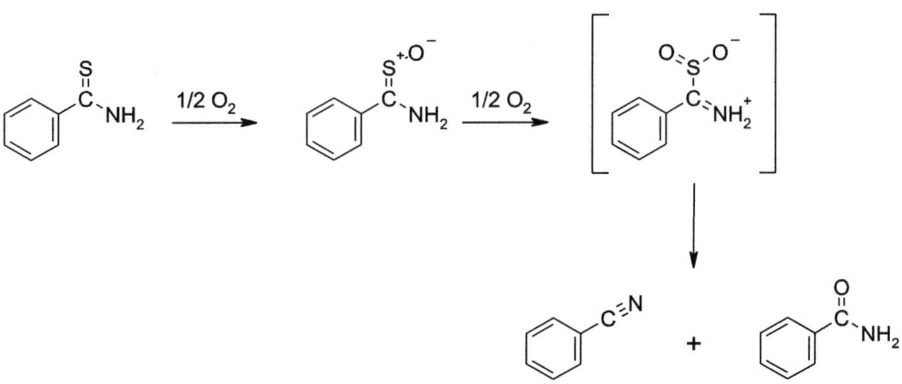

Fig. 4 Bacterial metabolism of thiobenzamide by oxygenases

7
Metabolic Rules for Each Functional Group

When new metabolism is elucidated, it can be characterized as a metabolic rule. Metabolic rules can be used to describe biodegradation in a generalizable way and further serve as the underlying basis of a system that can be used to predict metabolism for compounds not yet tested for biodegradation experimentally. A mixture of 60 functional groups containing two, three, four, or more groups in different combinations can theoretically generate an almost infinite array of organic molecules. The present constellation of over 9.3 million compounds in the Beilstein database on January 1, 2007, is large, but still small on the scale of what can and will be synthesized by organic chemists in coming decades. With biodegradation studies being much slower than organic synthesis of new chemicals, the gap will grow over time. The prediction of biodegradation metabolism will increasingly be needed to fill that gap.

The UM-BBD Pathway Prediction System (PPS) is based on a set of approximately 240 metabolic rules. The number is approximate because rules change, are added, or deleted, as knowledge grows. In general, the rules reflect the transformation of the 60 functional groups contained within the UM-BBD. There is an average of four metabolic rules for each functional group. A representative rule is shown in Fig. 5.

primary Alcohol ⟶ Aldehyde

Aerobic Likelihood: Likely

UM-BBD Reaction(s):

Benzyl alcohol ⟶ Benzaldehyde (reacID# r0267)

cis-3-Chloro-2-propene-1-ol ⟶ cis-3-Chloroallyl aldehyde (reacID# r0691)

.
.
.

Fig. 5 Representative bt rule for the oxidation of an alcohol to an aldehyde. Shown are only two of the over 30 UM-BBD reactions assigned to this rule. The complete rule, including all reactions, comments, and similarities, is available: http://umbbd.msi.umn.edu/servlets/rule.jsp?rule=bt0001

8
Predicting Biodegradation Based on Mechanistic Rules

Biodegradation prediction is important for many reasons, some of which have already been mentioned: to fill the gap between known chemical compounds and known metabolism; to help design environmentally safe molecules; to steer the chemical analysis in degradation route studies; and to know more about the fate of chemicals in the environment.

Regarding the latter, it is desirable to predict all metabolic pathways, to know which pathways might dead-end and accumulate products, and to discern between likely and unlikely pathways. By combining biodegradation route predictions with predictions of environmental mobility, ecotoxicity and human toxicity, described elsewhere in this book, we can more readily identify transformation products of most concern.

There have been numerous efforts to predict biodegradation metabolism and thus gain insight into the fate of chemicals in the environment. Systems include META [33], BESS [34], and CATABOL [35]. Although they all differ in design and implementation, they are generally based on expert knowledge and a "rule set" of some type. For example, the META system trains on clusters of atoms that frequently appear in metabolism and maps the transformation of that atom set into another atom set. This closely resembles a metabolic rule set based on functional groups.

The UM-BBD-PPS predicts all possible metabolic pathways based on its metabolic rules [10]. The user draws a structure or enters a SMILES (Simplified Molecular Input Line Entry System) string representing the compound of interest. The PPS identifies functional groups and matches them to appropriate rules. First-round metabolites are produced. Any first-round metabolite can be selected, and used to match another set of rules. When no metabolic rules are matched, the cycle is stopped. This could indicate a "non-metabolizable" compound or, in other cases, it is an "endpoint" metabolite that is a common intermediary compound.

Predicted pathways do not represent the metabolism of any single bacterium. By matching all possible rules, one obtains the set of meta-metabolic pathways. This reflects the reality that almost all environments are microbiologically complex. Terrestrial and aquatic environments have been known to contain a large range of microbial types, an observation supported by recent environmental genome studies. This, in turn, reflects a complexity of reaction types. Moreover, there are many studies illustrating that chemicals in the environment are often metabolized by metabolic sequences that span different microbes [36]. That is, a metabolite of one microbe can leave the cell and be acquired and further metabolized by another, completely different, bacterium. In this way, the PPS reflects this natural metabolism by microbial assemblages.

9
Combinatorial Explosion and Pathway Prioritization

It is desirable to reflect natural biodegradation diversity, but a complete set of all possible pathways can result in too many choices, a problem known as a "combinatorial explosion". For example, if there were ten metabolites at each stage of prediction and no convergence of metabolites, the number of metabolites would increase by an exponent of ten at each step. The resultant thousands of pathways would be beyond human evaluation on a reasonable time scale. To deal with this problem, it is necessary to further guide users by assigning priorities to every rule that governs each predicted reaction.

Prioritizing rules required additional expert knowledge and this was acquired by a series of workshops and consultation with biodegradation experts. Rules were prioritized on their likelihood of occurrence on a scale ranging from one to five. The numbers pertained to the following perceived likelihoods of a given reaction: (1) highly likely, (2) likely, (3) neutral, (4) unlikely, and (5) highly unlikely. The likelihood derived from expert evaluations for each rule that described a reaction occurring under defined standard conditions. The standard conditions were for an environment that was aerobic, 25 °C, standard pressure, adequate moisture, with no competing or toxic other reactions.

The UM-BBD-PPS was then able to provide the users with a predicted likelihood for each biotransformation reaction leading to a new metabolite. This gives the user an additional criterion for evaluating the generated metabolic pathways. A user may, for example, choose to only examine pathways that contain reactions being assigned likelihood values of 3 (neutral) or a correspondingly lower number (greater likelihood). Alternatively, a user may follow some pathways that have likely and very likely initial reactions followed by unlikely or highly unlikely reactions. The metabolite immediately preceding each unlikely reaction may turn out to be a compound that would accumulate in the environment. Thus, the fate of the putative accumulating compound may need to be considered for its human or ecosystem toxicity.

10
Usefulness and Future of Metabolite Predictions

The UM-BBD-PPS is designed to guide users, rather than offer predictions that appear absolute. This contrasts with the CATABOL prediction system that predicts only one metabolic pathway for a given compound and shows mass balances at any instant in time. While many users would like this precision, it is unlikely that only one pathway operates for most chemicals in nature, or that all environments are identical in biodegradative capability.

In this context, it is important to recognize that the biodegradation of organic compounds in the environment is very complex. Whitman et al. [37] have estimated that on the order of 10^{31} bacteria exist on earth. This represents a mass comparable to that of all the green plants on earth. Studies of extracted environmental DNA also show enormous species diversity in bacteria [38]. Perhaps 10^6 bacterial species were demonstrated in 1 g of typical soil. With such enormity and complexity, how can one reasonably improve on current biodegradation prediction?

It may be possible to overlay other knowledge on rule-based metabolite prediction. For example, analysis of complete pathways can reveal an overall biochemical logic, or perhaps the absence of a logical sequence. That is, microbes live under intense selective pressure, causing biodegradative pathways to be strongly selected with respect to efficiency. Efficiency in biological terms means that pathways that provide more energy and atoms for the cell will become more prevalent as the cells that carry them become more successful and outcompete other bacteria. Using this knowledge, we can predict pathways and then analyze them for overall thermodynamic efficiency. All pathways that transform a given organic compound to carbon dioxide with the same electron acceptor will have shown similar energy evolution. However, some pathways will capture more of the energy in the form of ATP or other energy currencies. We anticipate that high energy capture will be selected for over time, and so one could predict that such pathways are more likely, or more high priority in a prediction scheme. Such an analysis would then offer a PPS user another tool to assess the overall likelihood on one predicted pathway over another.

It is necessary to acknowledge our lack of knowledge of all biodegradation reactions, and the subtle influence of specific environmental conditions on biodegradation. In light of this, biodegradation prediction will never be perfect, but it is perfectible.

Acknowledgements We thank Chunhui Li for helpful discussions and help with the figures. This work was supported in part by National Science Foundation Grant NSF0543416 and Lhasa Limited.

References

1. Hogue J (2007) Chemical & Eng News 85:34
2. Brown MP, Bush B, Rhee GY, Shane L (1988) Science 240:1674
3. Abramowicz D (1994) Res Microbiol 145:42
4. Miller E, Wohlferth G, Diekert G (1998) Arch Microbiol 169:497
5. Krasotkina J, Walters T, Maruya KA, Ragsdale SW (2001) J Biol Chem 276:40991
6. Vogel TM, McCarty PL (1985) Appl Environ Microbiol 49:1080
7. Swaen GM, Duijts SF (2005) Epidemiologic evidence for the carcinogenicity of vinyl chloride monomer. Scand J Work Environ Health 31(3):233

8. Ellis LBM, Roe D, Wackett LP (2006) Nucl Acids Res 34:D517
9. Yamada H, Shimizu S, Kobayashi M (2001) Chem Rec 1:152
10. Hou BK, Wackett LP, Ellis LB (2003) J Chem Inf Comput Sci 43:1051
11. Di Lello P, Benison GC, Valafar H, Pitts KE, Summers AO, Legault P, Omichinski JG (2004) Biochemistry 43:8322
12. Grumping R, Michalke K, Hirner AV, Hensel R (1999) Appl Environ Microbiol 65:2276
13. Negrete-Raymond AC, Weder B, Wackett LP (2003) Appl Environ Microbiol 69:4263
14. Inoue H, Takimura O, Fuse H, Murakami K, Kamimura K, Yamaoka Y (2000) Appl Environ Microbiol 66:3492
15. Wackett LP, Hershberger CD (2001) Biocatalysis and Biodegradation: Microbial Transformation of Organic Compounds. American Society for Microbiology Press, Washington, DC
16. GOLD tables. GOLD, Genomes on-line. URL = http://www.genomesonline.org/gold.cgi
17. Ward N, Fraser CM (2005) Curr Opin Microbiol 8:564
18. Babbitt PC (2003) Curr Opin Chem Biol 7:23
19. Somerville CC, Nishino SF, Spain JC (1995) J Bacteriol 177:3837
20. Nishino SF, Spain JC (1993) Appl Environ Microbiol 59:2520
21. He Z, Nadeau LJ, Spain JC (2000) Eur J Biochem 267:1110
22. Davis JK, Paoli GC, He Z, Nadeau LJ, Somerville CC, Spain JC (2000) Appl Environ Microbiol 66:2965
23. Di Santo R, Costi R, Artico M, Massa S, Lampis G, Deidda D, Pompei R (1998) Bioorg Med Chem Lett 8:2931
24. Stover CK, Warrener P, VanDevanter DR, Sherman DR, Arain TM, Langhorne MH, Anderson SW, Towell JA, Yuan Y, McMurray DN, Kreiswirth BN, Barry CE, Baker (2000) Nature 405:962
25. Murugasu-Oei B, Dick T (2000) J Antimicrob Chemother 46:917
26. Fowden L (1956) Biochem J 64:323
27. Fowden L, Richmond MH (1963) Biochim Biophys Acta 71:459
28. Scarpulla RC, Soffer RL (1978) J Biol Chem 253:5997
29. Dunnill PM, Fowden L (1965) Phytochemistry 4:445
30. Burroughs AM, Allen KN, Dunaway-Mariano D, Aravind L (2006) J Mol Biol 361:1003
31. Hou BK, Ellis LBM, Wackett LP (2004) J Ind Microbiol Biotechnol 31:261
32. Dodge AG, Richman JR, Johnson G, Wackett LP (2006) Appl Environ Microbiol 72:7468
33. Klopman G, Tu M (1997) Environ Toxicol Chem 16:1829
34. Wackett LP, Ellis LBM, Speedie SM, Hershberger CD, Knackmuss H-J, Spormann AM, Walsh CT, Forney LJ, Punch WF, Kazic T, Kanehisa M, Berndt DJ (1999) ASM News 65:87
35. Jaworska J, Dimitrov S, Nikolova N, Mekenyan O (2002) SAR QSAR Environ Res 13:307
36. DeSouza ML, Newcombe D, Alvey S, Crowley DE, Hay A, Sadowsky MJ, Wackett LP (1998) Appl Environ Microbiol 64:178
37. Whitman WB, Coleman DC, Wiebe WJ (1998) Proc Nat Acad Sci USA 95:6578
38. Gans J, Wolinsky M, Dunbar J (2005) Science 309:1387

Hdb Env Chem Vol. 2, Part P (2009): 17–41
DOI 10.1007/698_2_012

Published online: 15 March 2008

Predicting the Persistence of Organic Compounds

Philip H. Howard

Syracuse Research Corporation, 7502 Round Pond Road, N. Syracuse, NY 13212, USA
howardp@syrres.com

Abstract Over 30 000 chemicals are used in commercial quantities and very few of these chemicals have experimental data on their environmental degradability/persistence. This chapter reviews databases which can be searched for persistence information and what to do when the chemical of interest does not have any data. Two general approaches are suggested: (1) identify chemicals that are similar in structure and have persistence data or (2) use general quantitative structure-degradation relationships (QSDR) models. It is concluded that estimation methods are available for the most important degradation processes: atmospheric oxidation, biodegradation, and hydrolysis.

Keywords Biodegradation · Persistence prediction · Structure/degradability relationships

Abbreviations

AOPWIN	Atmospheric oxidation rate prediction program
BIOWIN	Biodegradation prediction program
BOD	Biological oxygen demand

CAS	Chemical Abstracts Service
COD	Chemical oxygen demand
DOC	Dissolved organic carbon
EFDB	Environmental Fate Data Base
EPI	Estimation Programs Interface
HSDB	Hazardous Substances Data Bank
HYDROWIN	Hydrolysis rate prediction program
LFER	Linear free energy relationship
MCI	Molecular connectivity index
MITI	Ministry of International Trade and Industry
MO	Molecular orbital
OECD	Organisation for Economic Co-operation and Development
PBT	Persistent, bioaccumulative, and toxic
PLS	Partial least squares
QSAR	Quantitative structure-activity relationship
QSDR	Quantitative structure-degradability relationship
QSPR	Quantitative structure-property relationship
SRC	Syracuse Research Corporation
UM-BBD	University of Minnesota Biocatalysis/Biodegadation Database

1 Introduction

The need for predicting degradation/persistence assumes that experimental information on the chemical of interest is not available. In order to predict the degradation/persistence of organic chemicals in the environment when there are no measured data, either experimental data must be available on a close structural analog or the following two conditions must be met: (1) an estimation method must be available to predict the rate of degradation and (2) the method must model an important mechanism of degradation (biodegradation, atmospheric oxidation, chemical hydrolysis). This chapter will focus on the various approaches to predicting persistence of organic compounds in the absence of experimental information.

Of course, the best scenario in predicting persistence is the presence of relevant experimental data on the chemical of interest. However, experimental data for the approximately 30 000 chemicals that are used in commercial quantities [1] are rarely available. Sometimes, experimental data are available, but are not relevant to the degradation rate likely to be found in the environment. For example, many compounds have microbial pure culture studies, which are good for suggesting potential degradation pathways and the resulting transformation products, but provide little insight into degradation rates in soil or water [2]. This chapter will focus on the situation when no relevant data are available for the compound of interest.

If no data are available, an assessor can take two general approaches: (1) try to identify chemicals that are similar in structure, that do have data and use some general structure/"rules of thumb" to generate qualitative data on persistence [2] or (2) rely on general quantitative structure-degradation relationships (QSDR) models, many of which are available as computer programs. When large numbers of chemicals need to be evaluated, the latter approach is often the only feasible method. The structure/degradability "rules of thumb" have been known for many decades [2] and are easily understood (e.g., as the number of chlorines on an aromatic ring increases, so does the persistence).

2
Estimating Degradation from Experimental Data on Chemical Analogs

The process for identifying chemical analogs for degradation is very similar to the process for identifying chemical analogs for physical properties (see Fig. 1, reproduced with permission from the Society for Environmental Toxicology and Chemistry) [3]. The process begins with confirming the identity of the chemical whose degradation rate needs to be estimated. This chemical identification information usually includes the chemical name or synonym, Chemical Abstracts Service (CAS) Registry Number, and/or the chemical structure, all of which can be identified using a variety of databases, such as the Chemical Registry file of the American Chemical Society or free on-line files, such as ChemFinder and ChemID*plus* (Table 1).

Once the chemical identity is determined, the next step is to ascertain whether the chemical has any environmental degradation data. For this chapter, it will be assumed that literature searches will have already determined that no data on the chemical of interest are available. The approach and databases to be searched are similar to those used for substructure searching for analogs, as discussed in the next section.

3
Substructure Searching for Chemical Structure Analogs

If measured values are not available for the chemical of interest, a substructure search should be conducted to attempt to identify a close structural analog which has a measured value. Several options are available, a few of which allow the rapid identification of an analog with measured values. For example, there are free databases on the internet that are substructure searchable. ChemID*plus* (Table 1) is substructure searchable for all of the >6000 chemicals that are in the Hazardous Substances Data Bank (HSDB) as well as the 269 000 structures that are in the ChemID*plus* file. ChemS^3 (Table 1) can simultaneously substructure search the 20 000 chemicals in the four files of the Environmental Fate Data Base (EFDB) [4, 5].

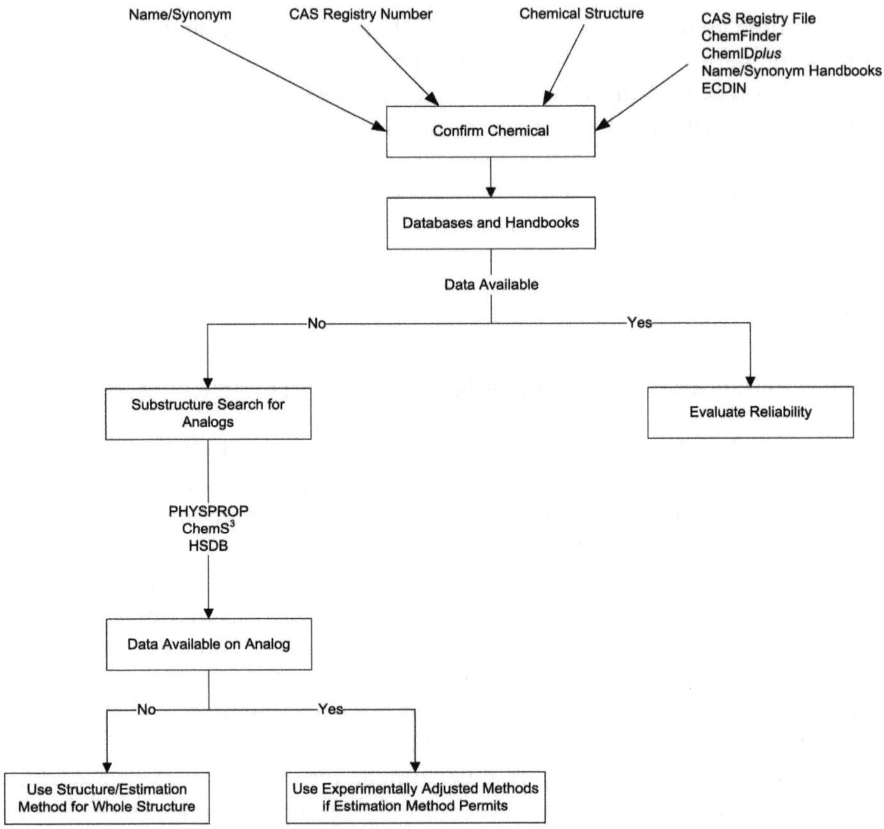

CAS = Chemical Abstracts Service; ECDIN = Environmental Chemicals Data and Information Network; HSDB = Hazardous Substances Data Bank; PHYSPROP = Physical Properties Database (contains ChemS^3, available online at http://esc.syrres.com/ChemS3/default.htm)

Fig. 1 Process for obtaining chemical property data

Once a structural analog with measured degradation/persistence properties is identified, there are some qualitative methods for adjusting the experimental value of the analog to obtain an estimate for the substance of interest. By analyzing differences between the analog and the target substance in terms of functional groups, one can determine in which direction (faster or slower) the degradation rate should change using the general effects of functional groups on persistence models. For example, if the property is biodegradation, fragments such as alcohols, acids, esters, and phenols should increase the biodegradability, while the addition of fragments such as halogens and nitro groups should decrease biodegradability [6, 7]. Also, one could determine if the differences in the structures would affect the biodegradation pathway (e.g., from the University of Minnesota Biocatalysis/Database

Table 1 Resources for chemical identification, degradation/persistence, and substructure searching

Chemical identification

Name	Number of chemicals	SS	Website/URL
CAS Registry File	30 818 220	√	http://stnweb.cas.org/
ChemFinder	78 000	√	http://www.chemfinder.com/
ChemID*plus*	367 438 (chemicals); 168 063 (structures)	√	http://chem.sis.nlm.nih.gov/chemidplus/

Databases of data

Name	Properties	Number of chemicals	Number of records	SS	Website/URL
Data resources with free web access					
ARS Pesticide Properties Database	Field dissipation, $ST^{1/2}$	334	334		http://www.ars.usda.gov/Services/docs.htm?docid=14199
US EPA REDs	Aquatic metabolism, soil metabolism, HS, PHOT	~ 1000	~ 1000		http://cfpub.epa.gov/oppref/rereg/status.cfm?show=rereg
CERIJ	Japanese MITI-I (OECD Test Guideline 301C) BD	≈ 900	≈ 900	√	http://www.safe.nite.go.jp/english/Haz_start.html http://www.safe.nite.go.jp/english/kizon/KIZON_start_hazkizon.html
SRC EFDB	BD, HS, PHOT, AO	27 478	473 060	√	http://esc.syrres.com/efdb.htm http://esc.syrres.com/pointer/default.asp
HSDB	OH, BD, HS, PHOT, terrestrial fate, aquatic fate	> 4500	> 4500	√	http://toxnet.nlm.nih.gov/ http://esc.syrres.com/pointer/default.asp
UM-BBD	163 microbial pathways, 1086 reactions	1026	NOS		http://umbbd.msi.umn.edu/

Table 1 (continued)

Name	Properties	Number of chemicals	Number of records	SS	Website/URL
NIST Chemistry WebBook	OH	50 762	NOS	√	http://webbook.nist.gov/chemistry
OSU Pesticide Properties Database	$ST^{1/2}$	341	341		http://ace.orst.edu/info/nptn/ppdmove.htm
PHYSPROP	OH	41 000	NOS	√	http://esc.syrres.com/interkow/physdemo.htm http://esc.syrres.com/pointer/default.asp http://esc.syrres.com/ChemS3/default.htm
Chemical degradation/persistence software (free or free access)					
EPI Suite™	OH, HS, AO, BD				http://www.epa.gov/oppt/exposure/pubs/episuitedl.htm
CATABOL	BD			√	http://oasis-lmc.org/?section=software&swid=1
Online log K_{OW} *calculators (free)*					
SPARC	HS	NOS	NOS	√	http://ibmlc2.chem.uga.edu/sparc/
UM-BBD: Pathway Prediction Results	BD			√	http://umbbd.msi.umn.edu/predict/
Resources that specialize in providing links to other resources					
ChemFinder					http://www.chemfinder.com/
ChemID*plus*					http://chem.sis.nlm.nih.gov/chemidplus/
ILPI MSDS					http://www.ilpi.com/msds/
NPIC					http://npic.orst.edu/

AO = atmospheric oxidation; ARS = Agricultural Research Service; BD = biodegradation; CERIJ = Chemical Evaluation and Research Institute, Japan; HS = hydrolysis; ILPI = Interactive Learning Paradigms, Incorporated; MSDS = Material Safety Data Sheet; NIST = National Institute of Standards and Technology; NOS = not otherwise specified; NPIC = National Pesticide Information Center; OH = hydroxyl radical rate constant; OSU = Oregon State University; PHOT = photolysis; PHYSPROP = Physical Properties Database; RED = Reregistration Eligibility Documents; SPARC = SPARC Performs Automated Reasoning in Chemistry; SS = substructure searching; $ST^{1/2}$ = soil half-life

[UM-BBD]). This kind of qualitative assessment will give a good indication of an approximate direction of persistence (increase or decrease) resulting from the presence of a functional group, and may also be useful for assessing the reliability of an estimated value based on the whole structure (these methods are reviewed in the next section).

Most experimental persistence data are usually from laboratory screening or grab sample tests; rarely are field studies available, unless the chemical is used as a pesticide. In some instances, monitoring data will be available, but it is very difficult to obtain degradation rates from monitoring data because of the many processes taking place (e.g., transport processes which do not degrade the chemical). The vast majority of data for non-pesticide chemicals is from Organisation for Economic Co-operation and Development (OECD) ready or inherent biodegradation tests [8]. However, most models that attempt to simulate the behavior of and exposure to a chemical in the environment need degradation rates or half-lives. The U.S. Environmental Protection Agency (EPA) has suggested a scheme for converting percent theoretical biological oxygen demand (BOD), chemical oxygen demand (COD), dissolved organic carbon (DOC) consumption, or CO_2 emission data (typical of what is reported in an OECD ready or inherent test) to rates and half-lives (Table 2). Using these rates for the chemical analog and adjusting them for the structural differences, as indicated above, should allow for semiquantitative estimates of degradation for the chemical of interest.

Table 2 Using ready and inherent biodegradability data to derive input data for the EQuilibrium Criterion (EQC) model [35]. Multimedia fate models like the EQC model require compartmental half-lives for air, water, soil, and sediment. The scheme in the table is offered as an interim procedure for assigning half-lives for input to such models. These are bulk half-lives (i.e., for the compartment as a whole). They are not to be interpreted as (necessarily) half-lives for any specific process, such as biodegradation. No assumptions which compromise their interpretation as bulk half-lives should be made, including, for example, the assumption that biodegradation is the important process and occurs in soil pore water only. Air half-lives are not addressed here and it is assumed that data for input to models are either measured or derived from AOPWIN or similar methodology.

Proposed scheme Ready test result	Inherent test result	Water half-life, days	Rate constant
Pass test	–	5	0.14 day^{-1}
No pass, but $\geq$ 40%	–	10	0.069 day^{-1}
No pass: $\geq$ 20 but < 40%	$\geq$ 70%	30	0.023 day^{-1}
–	$\geq$ 20 but < 70%	100	0.0069 day^{-1}
No pass: < 20%	< 20%	10 000, or other default for no biodegradation as appropriate	($k = 0$)

As an example, assume that a half-life is needed for 4-nitrosalicylic acid (4-nitro-2-hydroxylbenzoic acid). There are no biodegradation data for this compound in EFDB or HSDB (Table 1), but there is a considerable amount of data for salicylic acid, including one soil study with radiolabelled ^{14}C and an 88.1% for theoretical BOD in two weeks in the Japanese Ministry of International Trade and Industry (MITI) test (from the MITI test result, Table 2 would suggest a half-life of five days). All of the studies suggest that salicylic acid biodegradation rates will be very fast. However, adding a nitro group to an aromatic ring decreases the biodegradability, so the 4-nitrosalicylic acid biodegradation half-life will be considerably slower than salicylic acid. No exact value can be given, but something over 30 days seems reasonable by expert judgment. Since nitrosalicyclic acid has one hydroxyl and one carboxylic acid functional group, both of which increase biodegradability, the compound will not be extremely persistent (probably not over 60 days). For more input to the estimate, biodegradation QSDRs discussed below should be run for comparison.

4 Basics of Quantitative Structure-Degradation Relationships

All developments of quantitative structure activity relationships (QSARs)/quantitative structure-property relationships (QSPRs)/QSDRs go through similar steps: (1) collection of a database of measured values for model development and validation/evaluation, (2) selection of chemical descriptors (can include connection indices, atom, bond, or functional groups, molecular orbital calculations), (3) development of the model (develop a correlation between the chemical descriptors and the activity/property/degradation values) using a variety of statistical approaches (linear and non-linear regression, neural networks, partial least squares (PLS), etc. [9]), and (4) validate/evaluate the model for predictability (usually try to use a separate set of chemicals other than the ones used to train the model; external validation) [10].

Once the database of values is collected, the chemical descriptors need to be selected. The chemical descriptors can vary considerably and are often selected by what the investigator favors. For example, relationships developed by Kier and Hall and coworkers [11, 12] usually use molecular connectivity indices (MCIs) and electrotopological state indices. EPA/Syracuse Research Corporation (SRC) Estimation Programs Interface (EPI) Suite™ models usually use atom/functional group descriptors which are hand selected [13]. Some groups use computer software to select atom/functional groups (e.g., MultiCASE [14]), some use defined functional groups that have degradation pathway rules (e.g., CATABOL software [15]), and others use molecular orbital approaches. Each of these types of descriptors has advantages and disadvantages. For example, computer programs easily calculate overall MCI numbers to correlate to the persistence value, but the MCI numbers do not provide a mechanistic understanding of the degradation process and it is difficult to understand how changes in the structure affect the results (one exception is that some indices correlate with branching). Fragment/atom approaches are easily calculated and can provide an understanding of the influence of parts of the molecule and the calculations are very transparent. Molecular orbital (MO) calculations are derived from first principles (possibly more theoretically based) and calculate values for the whole molecule (three-dimensional analyses). However, before the calculations can be done, the molecule needs to be energy minimized and it is difficult to determine which structure is critical to the degradation rate (e.g., should the vapor phase or water solvated form be used to simulate when an enzyme is approaching?).

Once the descriptors have been selected, investigators need to select the statistical approach for developing the QSAR model. This can involve a number of techniques, such as multiple linear regression, partial least squares analysis, neural networks, and a variety of others [9]. These techniques need to be applied to both the training set (model development) and the validation set (assessment of predictability).

5
Available Quantitative Structure-Degradation Relationships

5.1
Biodegradation

Because of the importance of biodegradation to the determination of persistence, there are many QSDRs and computer programs available for calculating biodegradation. These have been recently reviewed in considerable detail [2, 16], and only the methods that are commonly used, that cover structurally diverse chemicals, and are well documented, trained, and validated on large numbers of chemicals will be reviewed.

5.1.1
Biodegradability Probability Program

One of the more popular group contribution QSDRs is a series of models/ collectively referred to as biodegradability probability program (BIOWIN), which consists of six aerobic biodegradation models [6, 7, 17] and one anaerobic model which has been recently released [18]. These models are available for free from U.S. EPA as part of the EPI Suite™ (Table 1).

The original model [17] contained 35 structural fragments whose coefficients were developed by linear and nonlinear regression using an experiment database (BIODEG file, available at http://www.syrres.com/esc/biodeg.htm [19]) of weight-of-evidence evaluations for 264 chemicals. Two years later, the descriptors used in the first two models were slightly revised [6] and semiquantitative estimates of rates of primary and ultimate biodegradation, gathered from a survey of experts who analyzed 200 carefully selected substances, were used to develop two new models (total of four models). Over the years, the linear and nonlinear BIOWIN probability models have come into fairly widespread use in chemical screening activities. The survey models have been used for multimedia modeling to identify substances that have PBT (Persistent, Bioaccumulative, and Toxic) properties (e.g., the PBT Profiler, available at http://www.pbtprofiler.net/).

The linear and nonlinear BIOWIN probability models were reparameterized for the MITI data [7] and, with the linear and non-linear MITI models, were added to BIOWIN for a total of six models. The training set used to develop the MITI models consisted of results (pass/no pass) from the OECD 301C MITI test for 884 discrete organic substances. Four new fragments – hydrazine, organotin, quaternary ammonium, and fluorine (-F) – were included, and some old fragments were modified. The most important were the modifications made to better account for molecular size and branching in alkyl-containing molecules and aromatics. Generally, the models predict separate validation sets with 80–90% accuracy. The coefficients for all six of the aerobic and the one anaerobic model are included in Table 3, and a sample output from BIOWIN 4.1 with the included anaerobic model is presented in Fig. 2.

Environmental half-life data were collected on over 200 chemicals and the values were compared with the calculated BIOWIN results [20]. Many of these chemicals had very large variability in half-lives, especially for more persistent chemicals, which makes them very difficult to predict. Quantitatively, the correlations were very poor, but, when binning was used (fast, slow), the predictions were reasonably accurate (70% or greater).

5.1.2 CATABOL

The CATABOL program [21] (available at http://oasis-lmc.org/?section=software&swid=1) takes a unique approach to predicting biodegradability with the assumption that if a metabolism pathway is available for a chemical, it will be biodegradable. The probability of each occuring biotransformation pathway is calibrated by using overall BOD and/or extent of CO_2 production databases. It is novel in that the biodegradation extent is assessed based on the entire pathway and not just the parent structure. It is also able to predict persistent intermediates and it considers effects of adjacent fragments. There are two types of transformations: spontaneous (can include chemical hydrolysis) and catabolic (only biotic processes). The hierarchy of transformations

Table 3 BIOWIN structural fragments and coefficients

	Linear model BIOWIN1	Non-linear model BIOWIN 2	Ultimate BIOWIN 3	Primary BIOWIN 4	MITI Linear BIOWIN 5	MITI Non-linear BIOWIN 6	Anaerobic BIOWIN 7
Nitroso [–N–N=O]	- 0.52448	- 3.2587	- 0.38513	0.01848	- 0.204532	- 12.23596433	-
Linear C4 terminal chain [CCC–CH_3]	0.10843	1.8437	0.29834	0.26907	-	-	- 0.317727891
Aliphatic alcohol [–OH]	0.15873	1.1178	0.15997	0.12945	0.161139	1.00414794	0.132763785
Aromatic alcohol [–OH]	0.11581	0.9086	0.05638	0.03969	0.064226	0.48842336	0.080722447
Aliphatic acid [–C(=O)–OH]	0.07269	0.6431	0.364605	0.38557	0.181163	1.13459688	0.186772405
Aromatic acid [–C(=O)–OH]	0.17686	2.4224	0.08787	0.00775	0.37697	2.44492398	0.265573477
Aldehyde [–CHO]	0.28463	7.1804	0.02232	0.19664	0.411394	2.74360672	0.122566557
Ester [–C(=O)–O–C]	0.17418	4.0795	0.14021	0.22896	0.343735	2.44616254	0.171852097
Amide [–C(=O)–N or –C(=S)–N]	0.21015	2.6913	- 0.05421	0.20543	0.126629	0.88586757	- 0.56787549
Triazine ring (symmetric)	0.00953	- 5.7252	- 0.24586	- 0.05752	0.116818	- 9.30058608	- 0.078251716
Aliphatic chloride [–CL]	- 0.11139	- 1.8528	- 0.17318	- 0.10061	0.001088	- 0.63913703	- 0.01465836
Aromatic chloride [–CL]	- 0.18242	- 2.0155	- 0.2066	- 0.16534	0.006172	- 0.21914857	- 0.402272298
Aliphatic bromide [–Br]	- 0.04617	- 4.4432	0.02895	0.03538	0.096749	- 0.55610275	0.359082642
Aromatic bromide [–Br]	- 0.11034	- 1.6779	- 0.136	- 0.15351	0.166778	1.50213475	-
Aromatic iodide [–I]	- 0.75862	- 10.0033	- 0.04494	- 0.12707	- 0.384025	- 12.52237701	-
Aromatic fluoride [–F]	- 0.80999	- 10.5318	- 0.40694	0.01346	-	-	-
Carbon with four single bonds and no hydrogens	- 0.18393	- 1.7232	- 0.21212	- 0.15344	0.067617	0.39898879	- 0.334230083
Aromatic nitro [–NO_2]	- 0.30504	- 2.5086	- 0.16959	- 0.10838	- 0.18759	- 2.40346978	- 0.214065567

Table 3 (continued)

	Linear model BIOWIN1	Non-linear model BIOWIN 2	Ultimate BIOWIN 3	Primary BIOWIN 4	MITI Linear BIOWIN 5	MITI Non-linear BIOWIN 6	Anaerobic BIOWIN 7
Aliphatic amine [–NH_2 or –NH–]	0.15383	1.1099	0.02444	0.04328	0.033286	– 0.28453537	0.177289503
Aromatic amine [–NH_2 or –NH–]	– 0.23375	– 1.907	– 0.13495	– 0.10838	– 0.157691	– 1.22637038	– 0.277817196
Cyanide/nitriles [–C#N]	0.307	4.644	– 0.08238	– 0.0652	0.071654	0.23395096	–
Sulfonic acid/salt → aromatic attach	– 0.22377	– 1.0283	0.14221	0.02162	0.022126	0.67802696	– 0.376831194
Sulfonic acid/salt → aliphatic attach	0.10837	6.8331	0.19259	0.17714	–	–	–
Polyaromatic hydrocarbon (four or more rings)	– 0.6573	– 10.1644	– 0.79934	– 0.70224	–	–	–
Pyridine ring	– 0.15457	– 1.6381	– 0.21417	– 0.01874	– 0.033494	– 0.45988648	0.641066312
Aromatic ether [–O-aromatic carbon]	0.13191	2.2483	– 0.05812	0.07712	0.19523	1.32268407	0.178015655
Aliphatic ether [C–O–C]	– 0.34736	– 3.4294	– 0.00867	– 0.00974	0.00147	– 0.10714727	– 0.257250884
Ketone [–C–C(=O)–C–]	0.00683	– 0.453	– 0.02248	– 0.02222	0.117739	0.83343688	– 0.391891246
Tertiary amine	– 0.20526	– 2.2229	– 0.2548	– 0.288	– 0.084833	– 0.83964945	– 1.074854205
Phosphate ester	0.31394	44.4087	0.15373	0.46535	0.154711	1.13052493	0.527009437
Alkyl substituent on aromatic ring	0.05467	0.5771	– 0.07485	– 0.06853	–	–	– 0.11445193
Azo group [–N=N–]	– 0.24183	– 8.2194	– 0.30036	– 0.05279	– 0.045873	– 10.61291841	–

Table 3 (continued)

	Linear model BIOWIN1	Non-linear model BIOWIN 2	Ultimate BIOWIN 3	Primary BIOWIN 4	MITI Linear BIOWIN 5	MITI Non-linear BIOWIN 6	Anaerobic BIOWIN 7
Carbamate or thiocarbamate	0.07954	1.0094	- 0.04671	0.19363	- 0.043478	0.4189022	–
Trifluoromethyl group [–CF_3]	- 0.52042	- 5.6696	- 0.51296	- 0.2744	–	–	–
Unsubstituted aromatic (3 or less rings)	0.3192	7.1908	- 0.58591	- 0.3428	–	–	- 0.263503551
Unsubstituted phenyl group (C_6H_5–)	0.12809	1.7991	0.02201	0.00489	–	–	0.21818207
Fluorine [–F]	–	–	–	–	0.017378	- 3.98784413	–
Aromatic-CH_3	–	–	–	–	0.041461	0.30720473	- 0.257320057
Aromatic-CH_2	–	–	–	–	- 0.055696	- 0.12459317	- 0.00733586
Aromatic-CH	–	–	–	–	- 0.009754	0.26242161	0.033086588
Aromatic-H	–	–	–	–	0.008218	0.12014128	- 0.095430138
Methyl [–CH_3]	–	–	–	–	0.000411	0.01942827	- 0.079572183
–CH_2– [linear]	–	–	–	–	0.049416	0.42949426	0.025989832
–CH– [linear]	–	–	–	–	- 0.050672	- 0.09977022	- 0.165850299
–CH_2– [cyclic]	–	–	–	–	0.019727	0.2365247	- 0.120013553
–CH– [cyclic]	–	–	–	–	0.012444	- 0.12945411	0.039450559
–C=CH [alkenyl hydrogen]	–	–	–	–	0.006189	0.02851468	- 0.073523308
Hydrazine [–N–NH–]	–	–	–	–	- 0.372979	- 14.6593076	–
Quaternary amine	–	–	–	–	- 0.009261	0.25503385	- 0.437702788
Tin [Sn]	–	–	–	–	0.132328	- 9.73860109	–

Table 3 (continued)

	Linear model BIOWIN1	Non-linear model BIOWIN 2	Ultimate BIOWIN 3	Primary BIOWIN 4	MITI Linear BIOWIN 5	MITI Non-linear BIOWIN 6	Anaerobic BIOWIN 7
Molecular weight parameter	– 0.00047607	– 0.0142	– 0.00220987	– 0.001442756	– 0.002975	– 0.02886875	–
Equation constant	0.74754581	3.0087	3.19917051	3.847737	0.712141	2.52565623	0.83608437589

```
SMILES : CLc1ccc(N(=O)(=O))cc1C(=O)OCC
CHEM   :
MOL FOR: C9 H8 CL1 N1 O4
MOL WT : 229.62
-------------------------- BIOWIN v4.10 Results ---------------------------

   Biowin1 (Linear Model Prediction)    :  Does Not Biodegrade Fast
   Biowin2 (Non-Linear Model Prediction):  Does Not Biodegrade Fast
   Biowin3 (Ultimate Biodegradation Timeframe):  Weeks-Months
   Biowin4 (Primary  Biodegradation Timeframe):  Days-Weeks
   Biowin5 (MITI Linear Model Prediction)    :  Not Readily Degradable
   Biowin6 (MITI Non-Linear Model Prediction):  Not Readily Degradable
   Biowin7 (Anaerobic Model Prediction):  Does Not Biodegrade Fast
   Ready Biodegradability Prediction:  NO

------+-----+--------------------------------------------+---------+---------
 TYPE | NUM |        Biowin1 FRAGMENT DESCRIPTION        |  COEFF  |  VALUE
------+-----+--------------------------------------------+---------+---------
 Frag |  1  |  Ester    [-C(=O)-O-C]                     |  0.1742 |  0.1742
 Frag |  1  |  Aromatic chloride    [-CL]                | -0.1824 | -0.1824
 Frag |  1  |  Aromatic nitro   [-NO2]                   | -0.3050 | -0.3050
 MolWt|  *  |  Molecular Weight Parameter                |         | -0.1093
 Const|  *  |  Equation Constant                         |         |  0.7475
============+============================================+=========+=========
   RESULT   |    Biowin1 (Linear Biodeg Probability)     |         |  0.3249
============+============================================+=========+=========

------+-----+--------------------------------------------+---------+---------
 TYPE | NUM |        Biowin2 FRAGMENT DESCRIPTION        |  COEFF  |  VALUE
------+-----+--------------------------------------------+---------+---------
 Frag |  1  |  Ester    [-C(=O)-O-C]                     |  4.0795 |  4.0795
 Frag |  1  |  Aromatic chloride    [-CL]                | -2.0155 | -2.0155
 Frag |  1  |  Aromatic nitro   [-NO2]                   | -2.5086 | -2.5086
 MolWt|  *  |  Molecular Weight Parameter                |         | -3.2606
============+============================================+=========+=========
   RESULT   |  Biowin2 (Non-Linear Biodeg Probability)   |         |  0.3326
============+============================================+=========+=========

 A Probability Greater Than or Equal to 0.5 indicates --> Biodegrades Fast
 A Probability Less Than 0.5 indicates --> Does NOT Biodegrade Fast

------+-----+--------------------------------------------+---------+---------
 TYPE | NUM |        Biowin3 FRAGMENT DESCRIPTION        |  COEFF  |  VALUE
------+-----+--------------------------------------------+---------+---------
 Frag |  1  |  Ester    [-C(=O)-O-C]                     |  0.1402 |  0.1402
 Frag |  1  |  Aromatic chloride    [-CL]                | -0.2066 | -0.2066
 Frag |  1  |  Aromatic nitro   [-NO2]                   | -0.1696 | -0.1696
 MolWt|  *  |  Molecular Weight Parameter                |         | -0.5074
 Const|  *  |  Equation Constant                         |         |  3.1992
============+============================================+=========+=========
   RESULT   |  Biowin3 (Survey Model - Ultimate Biodeg)  |         |  2.4558
============+============================================+=========+=========

------+-----+--------------------------------------------+---------+---------
```

Fig. 2 BIOWIN, AOPWIN, and HYDROWIN output for ethyl 2-chloro-5-nitrobenzoate

```
TYPE | NUM |        Biowin4 FRAGMENT DESCRIPTION           |  COEFF  |  VALUE
------+-----+--------------------------------------------+---------+---------
Frag  |  1  | Ester    [-C(=O)-O-C]                        |  0.2290 |  0.2290
Frag  |  1  | Aromatic chloride   [-CL]                    | -0.1653 | -0.1653
Frag  |  1  | Aromatic nitro  [-NO2]                       | -0.1084 | -0.1084
MolWt|  *  | Molecular Weight Parameter                   |         | -0.3313
Const|  *  | Equation Constant                            |         |  3.8477
============+============================================+=========+=========
  RESULT    |  Biowin4 (Survey Model - Primary Biodeg)  |         |  3.4717
============+============================================+=========+=========

Result Classification:   5.00 -> hours     4.00 -> days    3.00 -> weeks
 (Primary & Ultimate)    2.00 -> months    1.00 -> longer

------+-----+--------------------------------------------+---------+---------
TYPE | NUM |        Biowin5 FRAGMENT DESCRIPTION           |  COEFF  |  VALUE
------+-----+--------------------------------------------+---------+---------
Frag  |  1  | Ester    [-C(=O)-O-C]                        |  0.3437 |  0.3437
Frag  |  1  | Aromatic chloride   [-CL]                    |  0.0062 |  0.0062
Frag  |  1  | Aromatic nitro  [-NO2]                       | -0.1876 | -0.1876
Frag  |  3  | Aromatic-H                                   |  0.0082 |  0.0247
Frag  |  1  | Methyl  [-CH3]                               |  0.0004 |  0.0004
Frag  |  1  | -CH2-  [linear]                              |  0.0494 |  0.0494
MolWt|  *  | Molecular Weight Parameter                   |         | -0.6831
Const|  *  | Equation Constant                            |         |  0.7121
============+============================================+=========+=========
  RESULT    | Biowin5 (MITI Linear Biodeg Probability)  |         |  0.2658
============+============================================+=========+=========

------+-----+--------------------------------------------+---------+---------
TYPE | NUM |        Biowin6 FRAGMENT DESCRIPTION           |  COEFF  |  VALUE
------+-----+--------------------------------------------+---------+---------
Frag  |  1  | Ester    [-C(=O)-O-C]                        |  2.4462 |  2.4462
Frag  |  1  | Aromatic chloride   [-CL]                    | -0.2191 | -0.2191
Frag  |  1  | Aromatic nitro  [-NO2]                       | -2.4035 | -2.4035
Frag  |  3  | Aromatic-H                                   |  0.1201 |  0.3604
Frag  |  1  | Methyl  [-CH3]                               |  0.0194 |  0.0194
Frag  |  1  | -CH2-  [linear]                              |  0.4295 |  0.4295
MolWt|  *  | Molecular Weight Parameter                   |         | -6.6289
============+============================================+=========+=========
  RESULT    |Biowin6 (MITI Non-Linear Biodeg Probability)|         |  0.0302
============+============================================+=========+=========

 A Probability Greater Than or Equal to 0.5 indicates --> Readily Degradable
 A Probability Less Than 0.5 indicates --> NOT Readily Degradable


------+-----+--------------------------------------------+---------+---------
TYPE | NUM |        Biowin7 FRAGMENT DESCRIPTION           |  COEFF  |  VALUE
------+-----+--------------------------------------------+---------+---------
Frag  |  1  | Ester    [-C(=O)-O-C]                        |  0.1719 |  0.1719
Frag  |  1  | Aromatic chloride   [-CL]                    | -0.4023 | -0.4023
Frag  |  1  | Aromatic nitro  [-NO2]                       | -0.2141 | -0.2141
Frag  |  3  | Aromatic-H                                   | -0.0954 | -0.2863
```

Fig. 2 (continued)

```
 Frag |  1  |  Methyl  [-CH3]                            | -0.0796 | -0.0796
 Frag |  1  |  -CH2-  [linear]                           |  0.0260 |  0.0260
 Const|  *  |  Equation Constant                         |         |  0.8361
============+============================================+=========+=========
   RESULT   |   Biowin7 (Anaerobic Linear Biodeg Prob)   |         |  0.0517
============+============================================+=========+=========

 A Probability Greater Than or Equal to 0.5 indicates --> Biodegrades Fast
 A Probability Less Than 0.5 indicates --> Does NOT Biodegrade Fast
```

AOPWIN 1.92

```
SMILES : CLc1ccc(N(=O)(=O))cc1C(=O)OCC
CHEM   :
MOL FOR: C9 H8 CL1 N1 O4
MOL WT : 229.62
------------------- SUMMARY (AOP v1.92): HYDROXYL RADICALS -------------------
Hydrogen Abstraction       =   1.6617 E-12 cm3/molecule-sec
Reaction with N, S and -OH =   0.0000 E-12 cm3/molecule-sec
Addition to Triple Bonds   =   0.0000 E-12 cm3/molecule-sec
Addition to Olefinic Bonds =   0.0000 E-12 cm3/molecule-sec
Addition to Aromatic Rings =   0.0554 E-12 cm3/molecule-sec
Addition to Fused Rings    =   0.0000 E-12 cm3/molecule-sec

   OVERALL OH Rate Constant =   1.7171 E-12 cm3/molecule-sec
   HALF-LIFE =      6.229 Days (12-hr day; 1.5E6 OH/cm3)
   HALF-LIFE =     74.749 Hrs
------------------- SUMMARY (AOP v1.91): OZONE REACTION ----------------------

               ******  NO OZONE REACTION ESTIMATION ******
               (ONLY Olefins and Acetylenes are Estimated)

Experimental Database:  NO Structure Matches


Hydrogen Abstraction Calculation:
 Ksec  = 0.934 F(-CH3)F(-OC(=O)R)=0.934(1.000)(1.600)= 1.494
 Kprim = 0.136 F(-CH2-)=0.136(1.230)= 0.167
 H Abstraction TOTAL = 1.662 E-12 cm3/molecule-sec


OH Addition to Aromatic Rings Calculation:
Es+ = sp+(-CL) + sm+(-NO2) + sm+(-C(=O)-OCH2CH3) +  = 1.154
Es+ = sm+(-CL) + sp+(-NO2) + sp+(-C(=O)-OCH2CH3) +  = 1.671
Es+ = sm+(-CL) + sp+(-NO2) + sp+(-C(=O)-OCH2CH3) +  = 1.671
 Most negative Es+ = 1.154
 Log Kar = -11.71 - 1.34(Es+) cm3/molecule-sec
  Ring #1 Kar = 0.0554 E-12 cm3/molecule-sec
 TOTAL Kar =  0.0554 E-12 cm3/molecule-sec
```

HYDROWIN 2.00

```
SMILES : CLc1ccc(N(=O)(=O))cc1C(=O)OCC
CHEM   :
```

Fig. 2 (continued)

```
MOL FOR: C9 H8 CL1 N1 O4
MOL WT : 229.62
-------------------------- HYDROWIN v1.90 Results --------------------------


NOTE: Fragment(s) on this compound are NOT available from the fragment
    library. Substitute(s) have been used!!!  Substitute R1, R2, R3,
    or R4 fragments are marked with double astericks "**".

ESTER:   R1-C(=O)-O-R2                ** R1: -Phenyl [1 frag]
                                               meta: -NO2
                                         R2: -CH2-CH3
 NOTE: Ortho-position fragments(s) on Phenyl ring(s) are NOT CONSIDERED!!
 Kb hydrolysis at atom # 11:  1.352E+000  L/mol-sec

 Total Kb for pH > 8 at 25 deg C :  1.352E+000  L/mol-sec
 Kb Half-Life at pH 8:        5.933 days
 Kb Half-Life at pH 7:       59.331  days
```

Fig. 2 (continued)

is set according to descending probabilities of individual transformations. The CATABOL program is able to predict BOD/CO_2 with about 90% accuracy [16]. In Fig. 3, the degradation pathway along with the probability of each step is presented for the same sample chemical used with BIOWIN in Fig. 2.

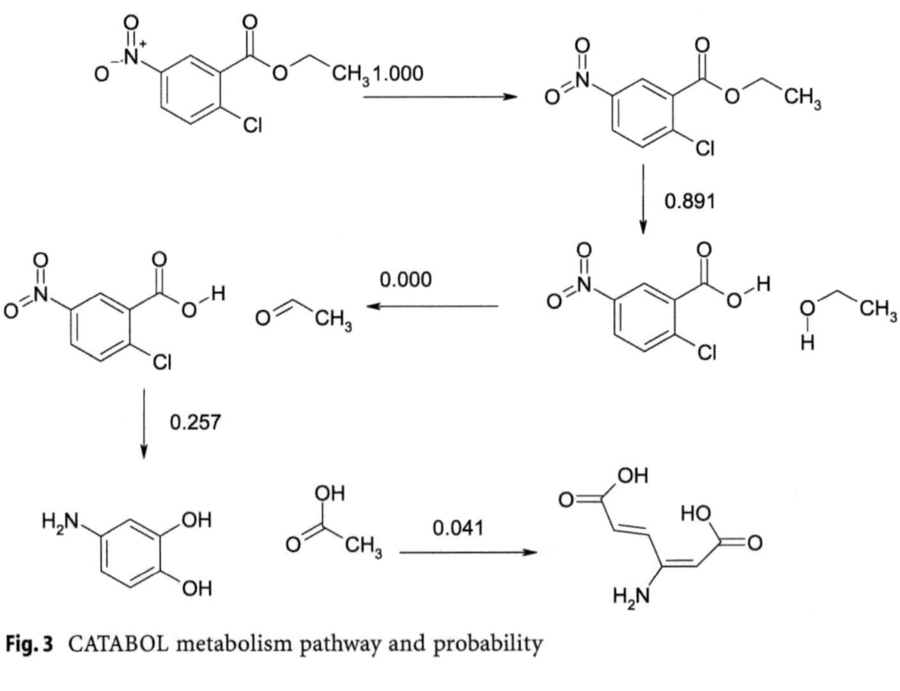

Fig. 3 CATABOL metabolism pathway and probability

5.1.3
UM-BBD Pathway Prediction System

The UM-BBD Pathway Prediction System, which is described in more detail in the chapter by Wackett and Ellis, is similar to CATABOL in that it gives pathways, but it is available on-line (Table 1) and does not give quantitative results. The rules are based upon pathways from the database, which are based on pure culture studies. Each rule for aerobic likelihood (takes place in moist soil or water at ambient conditions) is categorized as very likely, likely, neutral, unlikely, very unlikely, or unknown by two or more biodegradation experts; they are not statistically assigned, as is the case with CATABOL. This gives the user a sense of the likely pathways and possible persistent intermediates. The system does not give any numerical results from which biodegradation rates could be calculated. Some of the rules cover abiotic processes similar to CATABOL (e.g., acid chlorides to carboxylate; this is such a fast reaction that it must be chemical hydrolysis). In Fig. 4, the likely degradation metabolites are listed for the same sample chemical used with BIOWIN and CATABOL.

5.1.4
Use of Model Batteries to Increase Predictibility

It has been demonstrated that, by combining the results of more than one biodegradability model, the probability of the combined prediction is more accurate [22]. This requires that the performance statistics for each individual model are known.

6
Atmospheric Oxidation

In the atmosphere, there are three reactants that determine the rate of degradation of organics: hydroxyl radicals, ozone, and nitrate radicals. The hydroxyl radical is by far the most important, since it reacts with most organics, with the exception of fully halogenated compounds (no hydrogen to abstract). Ozone is important for only a small group of compounds, i.e., acetylenics and olefins. Nitrate radicals are only important at night and only react rapidly with a few classes of chemicals (e.g., phenols, mercaptans).

A variety of methods have been recently reviewed [23] and several new methods have been published since then [24]. There is one program, the atmospheric oxidation rate prediction program or AOPWIN – part of the EPI Suite™ software – that will calculate the hydroxyl radical and ozone rate constants and atmospheric half-lives (with selected oxidant concentrations) [23, 25]. The program uses the method of Atkinson and coworkers [26]

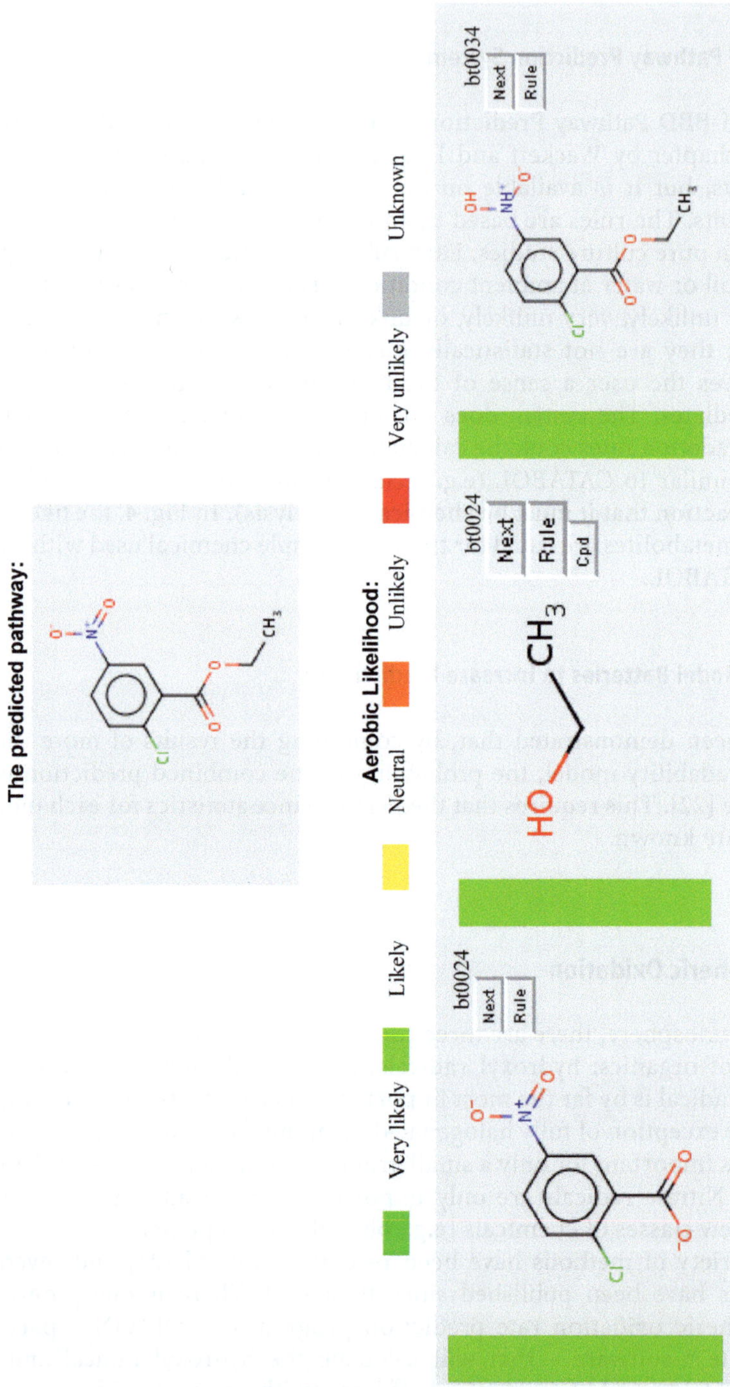

Fig. 4 UM-BBD: Pathway prediction results

and estimates hydroxyl radical rates by summing the following four pathways: H-atom abstraction from aliphatic C–H and O–H bonds, OH radical addition to olefinic (>C=C<) and acetylenic (–C≡C–) bonds, OH radical addition to aromatic rings, and OH radical reaction with selected nitrogen, sulphur, and phosphorus atom units. The estimation accuracy (logarithmic basis) for 720 chemicals is: correlation coefficient (r) = 0.977 ($r^2 = 0.955$), standard deviation = 0.246, and absolute mean error = 0.138. A sample output from AOPWIN 1.92 is presented in Fig. 2. The AOPWIN program will also calculate ozone rate constants using an adapted Atkinson and Carter [27] method. The estimation accuracy (logarithmic basis) for ozone for 112 alkenes and alkynes is: correlation coefficient (r) = 0.94 ($r^2 = 0.884$), standard deviation = 0.52, and absolute mean error = 0.35. Other ozone rate constant QSARs are reviewed in Meylan and Howard [23].

7 Chemical Hydrolysis

Chemical hydrolysis has been well studied by organic chemists for many years, frequently under environmentally relevant conditions (in water at ambient temperature and pH 5–9) [28]. Testing is sometimes done at elevated temperatures, and the rate constants are calculated at lower ambient temperatures using the Arrhenius equation. In some cases, solvent/water mixtures (e.g., 1% dioxane/water) are used to help dissolve very water-insoluble compounds. A recent review of existing estimation methods is available [29]. Several approaches to the QSARs have been used, including linear free energy relationship (LFERs) using Hammett and Taft constants (the most popular), correlations to pKa of the leaving group, and spectra (e.g., IR).

Chemical hydrolysis is only important for a limited number of chemicals which have functional groups that are susceptible to chemical hydrolysis under ambient environmental conditions. Most of these groups have been reviewed [28]. However, a number of new chemicals have been tested since then because of new requirements (e.g., pesticide registration, which requires testing at pH 5, 7, and 9) and these have some functional groups that are not covered in the review [28].

There are two programs that will calculate chemical hydrolysis rate constants. The SPARC on-line program (Table 1) uses a blend of polyparameter LFERs, QSAR, and perturbed MO methods [30, 31] to calculate the hydrolysis rates for carboxylic (see example in Fig. 4) and organophosphate esters. The other program is the hydrolysis rate prediction program or HYDROWIN, which is part of EPI Suite™. HYDROWIN uses linear free energy relationship (LFER) methodology [32] to estimate carboxylic esters, carbamates, epoxides, halomethanes, and selected alkyl halides (example in Fig. 2) and, in a soon-to-be-released version, organophosphate esters. Table 4 provides the

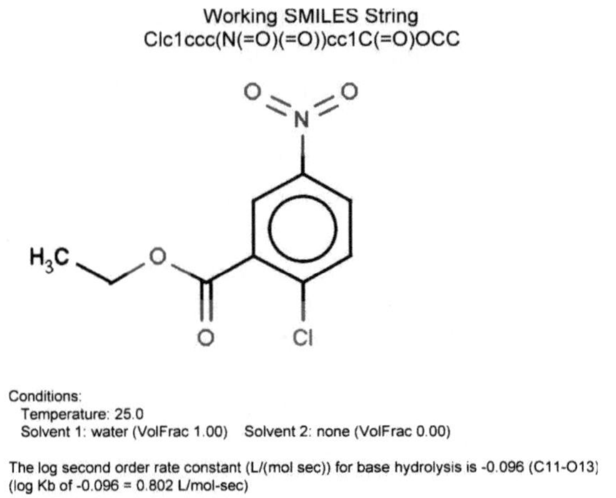

Fig. 5 Hydrolysis rate calculated by SPARC for esters

statistical correlation for the training set used for HYDROWIN as well as the LFER equations used for the predictions.

HYDROWIN is currently being updated to cover other chemical classes. However, because of the limited amount of experimental data, it is anticipated that most estimates for these classes will be only semiquantitative.

Table 4 HYDROWIN training set statistics and LFER equations

Chemical class	Number of chemicals and r^2	LFER equation
Esters	$n = 124$, $r^2 = 0.965$	$\log K_b$ = 0.92Es{R1} + 0.31Es{R2} + 2.16 sigma*{R1} + 2.30 sigma*{R2} + 2.10 sigmaX{R1} + 1.25 sigmaX{R2} + 2.67
Haloalkanes	$n = 7$, $r^2 = 0.98$	$\log K_b$ = 2.09 Summation[sigma*{R1+R2+R3}] + 0.491 Summation [Es{R1+R2+R3}] + 3.20 fx – 15.49
Halomethanes	$n = 12$, $r^2 = 0.996$	$\log K_b$ = 2.99 sigma*{Y2} + 2.83 Summation [Es{Y1+Y2+Y3}] + 0.995 fx – 0.633 fx is a halogen factor for the X substituent taken from the following list: F = 0.00, Cl = 1.33, Br = 2.60, I = 2.02
Aliphatic epoxides	$n = 14$, $r^2 = 0.80$	$\log K_a$ = 0.359 Summation[Es{R}] – 2.15 Summation[sigma*{R}] + 1.015 Co – 1.765
Vinyl epoxides	$n = 20$, $r^2 = 0.94$	$\log K_a$ =– 0.88 Summation[Es{R}] – 4.18 Summation[sigma*{R}] + 0.63CT + 0.47Do – 1.36 Co – 0.98

8 Other Processes

Unfortunately, other processes, such as photolysis and oxidation/reduction reactions, do not have any general QSARs. A review of photoreactions in surface waters indicates that only a small portion of synthetic organic compounds absorb ultraviolet light in the sunlight region (>295 nm) and then use that energy to photodegrade [33]. Knowing the ultraviolet spectrum will allow one to determine if direct photolysis is possible, but one still needs to know the efficiency of the reaction (quantum yield). If the ultraviolet spectrum is known, the compound absorbs light above 295 nm, and the quantum yield is known, the rates can be calculated using models such as EPA's GC-SOLAR model [33]. Chemicals in certain classes will react with photochemically-generated oxidants in water, but, because of the low concentrations of the oxidants in water, this is only important for a few chemical classes (e.g., phenols, furans, sulfides) [33].

The oxidation-reduction reactions in the environment have been reviewed [34]. Although the conditions leading to oxidation or reduction are created by living organisms, the redox reactions in natural systems may proceed without further mediation by organisms. A few QSARs are listed for oxidation and reduction reactions, but they are only for narrow chemical classes (phenols, nitrobenzenes, halocarbons) [34].

9 Conclusions

Estimation methods are available for the most important degradation processes: atmospheric oxidation, biodegradation, and hydrolysis. Therefore, it is possible to predict the persistence of organic compounds in the environment, even when diverse structural categories are being considered. Atmospheric oxidation rates for all organic chemicals can be calculated with considerable accuracy. Chemical hydrolysis rates for a limited set of chemical classes can also be calculated. Several comprehensive biodegradation methods which give semiquantitative results are available. These rely on either fragment constant or newly developed predicted pathway methodology.

References

1. Muir DCG, Howard PH (2006) Are there other persistent organic pollutants? Environ Sci Technol 40:7157
2. Howard PH (2000) Biodegradation. In: Boethling RS, Mackay D (eds) Handbook of property estimation method for chemicals: environmental and health sciences. CRC/Lewis Publishers, Boca Raton, FL, p 281

3. Boethling RS, Howard PH, Meylan WM (2004) Environ Toxicol Chem 23:2290
4. Howard PH, Sage GW, LaMacchia A, Colb A (1982) J Chem Inform Comput Sci 22:38
5. Howard PH, Hueber AE, Mulesky BC, Crisman JC, Meylan W, Crosbie E, Gray DA, Sage GW, Howard KP, LaMacchia A, Boethling R, Troast R (1986) Environ Toxicol Chem 5:977
6. Boethling RS, Howard PH, Meylan W, Stiteler W, Beauman J, Tirado N (1994) Environ Sci Technol 28:459
7. Tunkel J, Howard PH, Boethling RS, Stiteler W, Loonen H (2000) Environ Toxicol Chem 19:2478
8. OECD (Organisation for Economic Co-operation and Development) (2007) Chemicals testing – guidelines, section 3: degradation and accumulation. http://www.oecd.org/document/57/0,2340,en_2649_34377_2348921_1_1_1_1,00.html. Last visited: 12/7/07
9. Eriksson L, Jaworska J, Worth AP, Cronin MTD, McDowell RM, Gramatica P (2003) Environ Health Perspect 111:1361
10. Jaworska JS, Comber M, Auer C, Van Leeuwen CJ (2003a) Environ Health Perspect 111:1358
11. Kier LB, Hall LH (1976) Molecular connectivity in chemistry and drug research. Academic Press, New York
12. de Gregorio C, Kier LB, Hall LH (1998) J Comput Aided Mol Des 12:557
13. US EPA (US Environmental Protection Agency) (2007) EPI Suite™ Estimation Software. http://www.epa.gov/oppt/exposure/pubs/episuitedl.htm. Last visited: 12/7/07
14. Klopman G (1998) J Chem Inf Comput Sci 38:78
15. LMC (Laboratory of Mathematical Chemistry) (2007) CATABOL Software. Bourgas "Prof. As. Zlatarov" University, Bourgas Bulgaria. http://oasis-lmc.org/?section=software&swid=1. Last visited: 12/7/07
16. Jaworska JS, Boethling RS, Howard PH (2003b) Environ Toxicol Chem 22:1710
17. Howard PH, Boethling RS, Stiteler WM, Meylan WM, Hueber AE, Beauman JA, Larosche ME (1992) Environ Toxicol Chem 11:593
18. Meylan WM, Boethling RS, Aronson D, Howard PH, Tunkel J (2007) Environ Toxicol Chem 26:1785
19. Howard PH, Hueber AE, Boethling RS (1987) Environ Toxicol Chem 6:1
20. Aronson D, Boethling R, Howard P, Stiteler W (2006) Chemosphere 63:1953
21. Jaworska JS, Dimitrov S, Nikolova N, Mekenyan O (2002) SAR QSAR Environ Res 13:307
22. Boethling RS, Lynch DG, Jaworska JS, Tunkel JL, Thom GC, Webb S (2004) Environ Toxicol Chem 23:911
23. Meylan WM, Howard PH (2003) Environ Toxicol Chem 22:1724
24. Öberg T (2005) Atmosph Environ 39:2189
25. Meylan WM, Howard PH (1993) Chemosphere 26:2293
26. Atkinson R (2000) Atmospheric oxidation. In: Boethling RS, Mackay D (eds) Handbook of property estimation methods for chemicals: environmental and health sciences. CRC/Lewis Publishers, Boca Raton, FL, p 335
27. Atkinson R, Carter WPL (1984) Chem Rev 84:437
28. Mabey W, Mill T (1978) J Phys Chem Ref Data 7:383
29. Wolfe NL, Jeffers PM (2000) Hydrolysis. In: Boethling RS, Mackay D (eds) Handbook of property estimation method for chemicals: environmental and health sciences. CRC/Lewis Publishers, Boca Raton, FL, p 311
30. Hilal SH, Carreira LA, Karickhoff SW (1994) Estimation of chemical reactivity parameters and physical properties of organic molecules using SPARC. In: Politzer P,

Murray JS (eds) Quantitative treatments of solute/solvent interactions, theoretical and computational chemistry. Elsevier, Amsterdam, p 291
31. Whiteside TS, Hilal SH, Carreira LA (2005) QSAR Comb Sci 24:123
32. Mill T, Haag W, Penwell P, Pettit T, Johnson H (1987) Environmental fate and exposure studies. Development of a PC-SAR for hydrolysis: Esters, alkyl halides and epoxides. EPA Contract No. 68-02-4254. SRI International, Menlo Park, CA
33. Mill T (2000) Photoreactions in surface waters. In: Boethling RS, Mackay D (eds) Handbook of property estimation method for chemicals: environmental and health sciences. CRC/Lewis Publishers, Boca Raton, FL, p 355
34. Tratnyek PG, Macalady DL (2000) Oxidation-reduction reactions in the aquatic environment. In: Boethling RS, Mackay D (eds) Handbook of property estimation method for chemicals: environmental and health sciences. CRC/Lewis Publishers, Boca Raton, FL, p 383
35. US EPA (2000) Interim guidance for using ready and inherent biodegradability tests to derive input data for multimedia models and wastewater treatment plants (WWT) models (9/1/2000). http://www.epa.gov/oppt/exposure/pubs/halflife.htm. Last visited: 12/7/07

Hdb Env Chem Vol. 2, Part P (2009): 43–81
DOI 10.1007/698_2_016

Published online: 13 September 2008

Analyzing transformation products of synthetic chemicals

Sandra Pérez[1] · Mira Petrovic[1,2] · D. Barceló[1] (✉)

[1]Department of Environmental Chemistry, IIQAB, CSIC, c/ Jordi Girona 18–26, 08034 Barcelona, Spain
dbcqam@iiqab.csic.es

[2]ICREA – Catalan Institution for Research and Advance Studies, Passeig Lluis Companys 23, 08010 Barcelona, Spain

Abstract This chapter gives an overview of strategies used in the identification and analysis of environmental transformation products of three important groups of synthetic chemicals: pesticides, pharmaceuticals, and personal care products. The characteristics and features of modern mass spectrometric instrumentation coupled to liquid chromatographic separation techniques as well as complementary techniques are presented and examples of their application to the characterization of transformation products of synthetic chemicals are described. Analytical methodologies for the quantitative analysis of the intact parent compounds and their transformation products in the environment are compiled.

Keywords Biodegradation · Mass spectrometry · Photochemistry · Synthetic chemicals · Transformation products

Abbreviations

amu	Atomic mass unit
APCI	Atmospheric pressure chemical ionization
API	Atmospheric pressure ionization
CID	Collision-induced dissociation
DAD	Diode array detector
DCDD	2,7-Dichlorodibenzo-*p*-dioxin
DDD	Dichlorodiphenyldichloroethane

DDE	Dichlorodiphenyldichloroethylene
DDT	1,1,1-Trichloro-2,2-bis(*p*-chlorophenyl)ethane
EI	Electron impact ionization
EPI	Enhanced product ion
ESA	Ethane sulfonic acid
ESI	Electrospray ionization
FL	Fluorescence detector
GC	Gas chromatography
IR	Infrared
LC	Liquid chromatography
LIT	Linear ion trap
MAE	Microwave-assisted extraction
MALDI	Matrix-assisted laser desorption/ionization
MASE	Microwave-assisted Soxhlet extraction
MDL	Method detection limit
MIPs	Molecular imprinted polymers
MS	Mass spectrometry
MS/MS	Tandem mass spectrometry
MW	Molecular weight
m/z	Mass to charge
NMR	Nuclear magnetic resonance
OA	Oxanillic acid
OTC	Oxytetracycline
PLE	Pressurized liquid extraction
PPCPs	Personal care products
ppm	Parts per million
PTP	Phototransformation product
QqQ	Triple quadrupole
SPE	Solid-phase extraction
ToF	Time of flight
TPs	Transformation products
t_R	Retention time
UPLC	Ultra-performance liquid chromatography
UV	Ultraviolet
WWTP	Wastewater treatment plant

1 Introduction

Elucidation of degradation pathways and identification of transformation products (TPs) is of crucial importance in understanding their fate in the environment and requires the employment of advanced instrumental techniques. Analytical methods that can be used for this purpose include liquid chromatography with diode array or fluorescence detector (LC-DAD/FL), nuclear magnetic resonance (NMR), infrared spectroscopy (IR), matrix-assisted laser desorption/ionization–mass spectrometry (MALDI-MS), gas chromatography–mass spectrometry (GC-MS); liquid chromatography–mass spectrometry (LC-MS). LC-MS has gained popularity and become one of the preferred techniques for analyzing polar contaminants and TPs formed in the environment [1]. The main drawback of GC for the analysis of polar compounds is that this technique is only amenable to compounds with high vapor pressure. As most drugs bear functional groups that impede GC analysis they need to be derivatized prior to injection into the gas chromatograph. For that reason, the combination of atmospheric pressure ionization (API)-MS with separation techniques such as LC or ultra-performance liquid chromatography (UPLC) has become the method of choice in the analysis of small polar organic molecules.

Single quadrupole LC-MS offers good sensitivity but, when very complex matrices like raw sewage extracts are investigated, insufficient selectivity often impairs the unequivocal identification of the target analytes. Tandem MS (MS/MS) affords superior performance in terms of sensitivity and selectivity in comparison with single quadrupole instruments, as it enables isolation of the molecular ion of the compound of interest in the first stage of the mass analyzer. Liquid chromatographic techniques coupled to MS/MS or hybrid mass spectrometers with distinct analyzers such as triple quadrupole (QqQ), time-of-flight (ToF), quadrupole time-of-flight (QqToF), quadrupole ion trap (IT), and recently the quadrupole linear ion trap (QqLIT) are the most widely used instrumental techniques in the analysis of organic pollutants [2].

This chapter provides an overview of the analysis of TPs in the environment of three important classes of synthetic chemicals namely pesticides, human and veterinary pharmaceuticals, and personal care products (PPCPs). A series of analytical protocols applied to determine and analyze TPs of manmade chemicals originating from photolysis as well as from microbial degradation in the environment and wastewater treatment plants (WWTPs) is presented. Furthermore, strategies for identifying unknown TPs of xenobiotic compounds including pesticides and PPCPs are presented based on the combination of mass spectrometric techniques and NMR, IR, and optical detection systems like DAD and FL.

2 Techniques for Analyzing Transformation Products of Synthetic Chemicals

2.1 Qualitative Analysis

In this section, MS and other techniques used for identification of TPs are reviewed. Some representative examples of the application of selected techniques for TPs identification are also presented.

2.1.1
Mass Spectrometric Techniques

GC-MS is a method that combines the features of gas–liquid chromatography and mass spectrometry to identify and quantify volatile analytes. Most GC-MS instruments employ electron impact ionization (EI), a kinetically controlled ionization process, yielding spectra for a given compound that are nearly reproducible over time and between different GC instruments [3]. This has allowed the generation of a comprehensive spectral library for compound identification. GC-MS is suitable for the analysis of non-polar volatile compounds and of highly volatile compounds with low vapor pressures. But, since TPs of manmade compounds are usually polar compounds, they need to undergo time-consuming extraction and derivatization processes prior to GC-MS analysis. Consequently, LC-MS using the API interfaces, atmospheric pressure chemical ionization (APCI), or electrospray ionization (ESI) has considerably changed the analytical methods used to determine polar compounds in aqueous environmental samples. ESI has become the most important ionization techniques in mass spectrometry for the on-line coupling with LC in the analysis and identification of low molecular-mass molecules [4]. In single-quadrupole MS instruments, structural information of TPs can be obtained by inducing in-source fragmentation, but collision-induced dissociation (CID) in triple-quadrupole and IT instruments offers higher selectivity. By increasing the voltage between sample cone and the first quadrupole instrument, a CID spectrum may also be generated in single quadrupole. However, it lacks selectivity as coelution of analyte and matrix components may yield a complicated and ambiguous spectrum.

Tandem mass analyzers afford selectivity by mass separation at two stages. When sensitivity is an issue, QqQ instruments are a powerful alternative for detection of TPs (see Table 1).

IT-MS uses three electrodes to trap ions in a small volume. A mass spectrum is obtained by changing the electrode voltages to eject the ions from the trap. The advantages of IT mass spectrometers include compact size, relatively inexpensive instrumentation, the ability to trap and accumulate ions to increase the signal-to-noise ratio of a measurement, and they possess MS^n capabilities. The latter feature is particularly attractive for the identification of transformation products because sequential fragmentation allows one to propose fragmentation pathways that, in many cases, are not as obvious in product ion spectra generated on QqQ or QqToF instruments [5]. Due to their small trapping volume, however, IT-MS have a limited capacity for ion storage, and overfilling of the IT results in deterioration in the mass spectrum and loss of the dynamic response range due to space charging [2]. This phenomenon can become a critical factor when ion ratios serve as criteria for compound identification, as established in analytical guidelines by regulatory agencies. Whereas for QqQ instruments variations of ion ratios are typically

Table 1 Comparison of characteristics of different mass analyzers

	Quadrupole	Ion trap	Time-of-flight	FT-ICR	Magnetic sector
Mass range	++	++	++++	+++	+++
Resolution	+	++	+++	++++	+++
Dynamic range	+++	++	++	++	++++
Mass accuracy	+	+	+++	++++	++++
Cost	+	++	++	++++	++++
Advantages	• High selectivity and sensitivity in triple-quadrupole instruments	• Compact system • Multiple-stage MS • High sensitivity	• Fastest analyzer • High ion transmission	• Multiple-stage MS	• Very high repruducibility • Isotope ratio measurements

within 10–15%, the variability in IT-MS can be as high as 30%. Trapping of ions can also be performed in linear ion traps that have two major advantages over conventional ion traps: a larger ion storage capacity and a higher trapping efficiency. The QTRAP system is based on a triple quadrupole in which the third quadrupole (Q3) can be operated either as normal quadrupole or in the LIT mode. In the latter mode, the trapped ions are ejected axially in a mass-selective fashion and are detected by the standard detector of the system. As yet, QqLIT systems require considerably higher investment than QqQ instruments (see Table 1).

Further improvements in selectivity can be achieved by the use of high resolution mass analyzers including ToF and Fourier transform (FT) (see Table 1). ToF instruments measure the mass-dependent time it takes ions of different mass-to-charge ratios to move from the entrance of the analyzer, where they have been orthogonally accelerated in a pulsed fashion, to the detector. Full-scan sensitivity, high-mass resolution, and mass accuracy provided by ToF-MS are very attractive for identification of transformation products. Mass errors below 2 mDa or 5 ppm are achievable on modern ToF instruments to propose and confirm elemental compositions [6]. Even more powerful in terms of confirmatory analysis are hybrid QqToF-MS systems that allow MS^2 experiments to be performed to provide fragmentation patterns together with accurate mass measurements of product ions (precision in the low parts per million range) [7]. FT-MS is also a high-resolution analyzer, where masses can be determined with very high accuracy (see Table 1). FT-MS is one of the most sensitive ion detection methods with resolution typically ranging from 10^5–10^6 and a mass accuracy of < 1 ppm [8]. However, the high cost of instrumentation currently restricts the circle of potential costumers to laboratories in the pharmaceutical industry [9]. An alternative to the two classical high-resolution mass spectrometers (FT and magnetic sector) is the recently launched LTQ Orbitrap that combines a conventional LIT-MS with an Orbitrap mass analyzer. This system provides outstanding mass accuracy, mass resolution, and reliable high sensitivity MS^n performance.

Another ToF instrument used for analysis is MALDI-ToF-MS. MALDI is a soft-ionization that causes little or no fragmentation of analytes, allowing the molecular ions of analytes to be identified, even within mixtures. MALDI-ToF MS analysis is sensitive and very rapid as once the sample has been mixed with a matrix on a MALDI target, a spectrum can be generated within seconds. MALDI-ToF-MS has been successfully used for the analysis of a wide range of different macromolecules.

2.1.2
Complementary Techniques

In some instances, with MS alone it is not possible to distinguish isomers or identify which part of the molecule has undergone modification in the transformation of the parent compound. Prior to MS analysis, derivatization of functional groups or H/D-exchange experiments can be performed.

NMR and MS have played an invaluable role in the structural characterization and quantification of TPs [10]. NMR allows the unambiguous determination of a compound's structure and stereochemistry generating resonances from the two most commonly measured nuclei: ^{1}H (the most receptive isotope at natural abundance) and ^{13}C, although nuclei from isotopes of some heteroatoms (^{19}F, ^{31}P) can also be observed. However, stand-alone NMR lacks the ability to separate molecules, and its low sensitivity requires time-consuming sample preparation and considerable amounts of material (typically > 1 mg) [10]. Coupling LC to NMR mitigates the significant pre-analysis of the samples, but not the lack of sensitivity with respect to LC-MS [10].

Additional approaches using spectrophotometric detection techniques (UV and FL) in line with MS can help identify TPs in the environment. UV-spectra have been used for qualitative purposes for a long time. However, compared to other identification techniques (such as NMR and MS) the broad spectral bands and the relatively small frequency range in UV-spectrometry are not particularly informative. Beside compound identification, the UV trace can be used for quantification of interference-free analytes. The use of a diode array detector (DAD) allows one to collect full UV spectra of the analytes eluting from the LC column, and both identification and quantification can be done. DAD instruments are suitable as detectors in HPLC. DAD spectra are solvent-sensitive; comparing spectra from reversed phase runs in acetonitrile to reference spectra collected in methanol may not match due to a 1–2 nm shift for some compounds [11]. Spectra collected during solvent gradients may also suffer from this problem, and an even worse problem occurs due to different solvents in fluorescence detection [12]. The selectivity of FL detection is also very high, owing to the very low probability of coeluting compounds that coincide with the target analyte in excitation and emission wavelengths [12]. Radiation with frequencies between 4000 and 400 cm^{-1} can be utilized in organic structure determination by making use of the fact that it is absorbed by bonds in organic compounds. The frequencies at which absorptions of IR radiation occurs (peaks or signals) can be correlated directly to bonds within the compound in question [13]. Therefore, IR can be considered a complementary technique for compound identification.

2.1.3
Identification of Transformation Products

In this section, some representative examples of the identification of TPs are presented, showing the potential of combining the different mass analyzers or instrumental techniques reviewed in the previous sections. Next, nine representative examples of characterization of TPs of synthetic chemicals using advance MS and/or the combination of other techniques are presented, as well as derivatization or H/D exchange.

Whereas studies on the environmental photochemistry of the majority of pesticides have been conducted extensively, few data exist for PPCPs. Pharmaceuticals are mainly polar compounds containing acidic or basic functional groups (such as carboxylic acids, phenols, and amines) that may be subject to direct and indirect photolysis. Although microbial degradation in waters and soil has been reported for pesticides, less work is reported for PPCPs. The result of such processes can be a complex mixture of reactive intermediates and TPs. Their identification represents a more challenging task than the identification of transformation products stemming from microbial transformation, for which at least some common mechanisms are well established. Therefore, the application of advanced instrumental techniques is of crucial importance.

The combination of MALDI-ToF-MS, NMR, and UV detection was reported [14] for the characterization of four photoproducts of atorvastatin in water. This compound, a blood lipid regulator, is one of the most prescribed drugs in the USA and Europe [15]. These photoproducts comprise a lactam ring arising from oxidation of a pyrrole ring and an alkyl/aryl shift. For example, for the identification of the photoproduct 1, MALDI-MS was used. The ESI spectrum showed a molecular peak at m/z 575 suggesting, along with the elemental analysis, a molecular formula of $C_{33}H_{34}FN_2O_6$. The UV spectrum revealed a band at 203 nm, a shift of 3 nm from the UV spectra of the parent compound (atorvastatin, UV 206). The ^{1}H- and ^{13}C-NMR spectra of the photoproduct 1 and comparison with the spectral data for atorvastatin revealed the presence of the following functionalities: three carbonyl groups, one quaternary sp^3-carbon, three aliphatic methines (the first two bearing oxygen), four aliphatic methylene carbons, two methyls, two quaternary sp^2-carbons, two monosubstituted aromatic rings, and one disubstituted aromatic ring. These correlations were consistent with the structure of photoproduct 1 (the formation of a lactam ring and the fluor-substituted-aryl shift from the parent compound) [14].

The combination of LC single quadrupole MS, LC-IT-MS, NMR, and IR was used for the identification of the phototransformation products of a potent fluoroquinolone antimicrobial used in human medicine, clinafloxacin, in aqueous solutions at pH<4 [16] (see Table 2). The structures of eight new photodegradation products (see Table 3) were identified that were produced by two degradation routes: (i) dechlorination, followed by further reactions involving the quinolone ring and loss of chlorine, to yield four polar products; and (ii) degradation of the pyrrolidine side chain without loss of chlorine, yielding several non-polar photodegradation products. The photoproduct polar 2 (PTP-2) of clinafloxacin was identified by the combination of NMR, LC-MS, and LC-IT-MS in ESI(+) mode. Comparison of the ^{1}H-NMR spectra of PTP-2 and PD 124 979 (the des-chloro analog of clinafloxacin) showed that in the aromatic region (7–9 ppm) the proton in the eight position of PD 124 979 (where the chlorine is located in clinafloxacin; see Table 2) is absent in the spectrum of PTP-2. This suggests the formation of a bond from the cyclopropyl group to the eight position of the quinolone ring. The characteristic isotope ratio of the molecular ion in LC-ESI-MS showed that PTP-2 and the other polar degradation products formed did not contain chlorine [16]. To provide a reference data set, the fragmentation pathway of clinafloxacin was determined. The mass spectrum of clinafloxacin showed the formation of four main fragment ions corresponding to the loss of: (i) NH_3 m/z 349, (ii) H_2O m/z 348, (iii) CO_2 m/z 322, and (iv) H_2O and C_2H_5N m/z 305. The fragmentation pattern of PTP-2 was analogous to that seen for clinafloxacin (see Table 3) and supported the proposed structure with NMR technique [16]. The photoproduct non-polar 10 (PTP-10) of clinafloxacin was characterized by the combination of LC-MS in APCI(+) mode, NMR, and IR. The characteristic isotope ratio of the molecular ion in LC-MS showed that PTP-10 and the other non-polar compounds formed did not exhibit loss of chlorine [16]. Comparison of the ^{1}H-NMR spectra of PTP-10 and clinafloxacin showed key differences in the pyrrolidine region (2.0–2.3 and 3.6–3.9 ppm). A new set of signals appeared in the spectrum of PTP-10 at 2.6 and 3.9 ppm. The ^{13}C-NMR spectrum showed resonances at 165, 176, and 212 ppm that are indicative of carbonyl carbons. The IR spectrum displayed an intense band at 1759 cm^{-1} not seen in the spectrum of clinafloxacin, which is in the frequency range expected for the absorption by a cyclic ketone in a five-membered ring. Thus, the proposed structure was formed by the transformation of the pyrrolidine side chain in a five-membered ring cyclic ketone [16].

The degradation of a synthetic fluoroquinolone, danofloxacin, in 72 pure cultures of microbes commonly found in soil was investigated [17]. This synthetic compound is exclusively used for veterinary medicine and can con-

Table 2 Identified transformation products of PPCPs in waters

Pharmaceutical (therapeutic class)	Structure	Identified microbial degradates in the aquatic environment and in activated sludge	Identified photochemical degradates (environmental conditions)
Atorvastatin (blood lipid regulator)	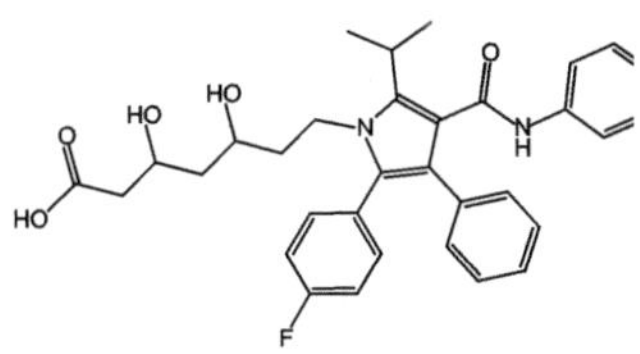	Not reported	4 TPs [14, 57]
Bezafibrate (blood lipid regulator)	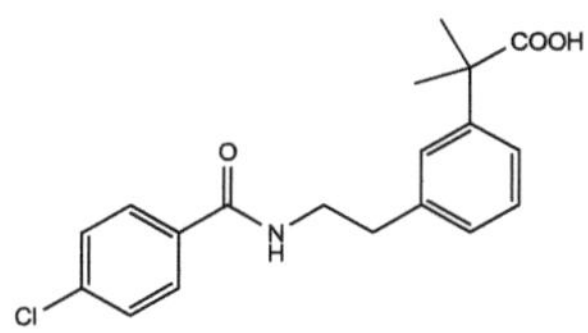	Activated sludge: 4-chlorobenzoic acid [23]	Not reported
Carbamazepine (psychiatric drug)	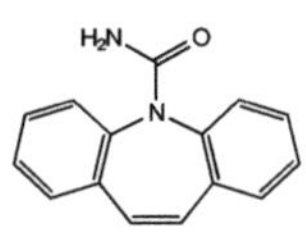	Not reported	5 TPs [58]
Chlortetracycline (antimicrobial)	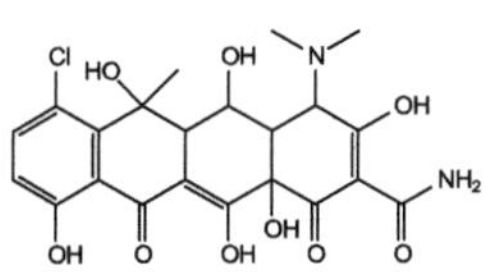	Not reported	One TP: Oxigenated chlortetracycline [59]
Citalopram (psychiatric drug)	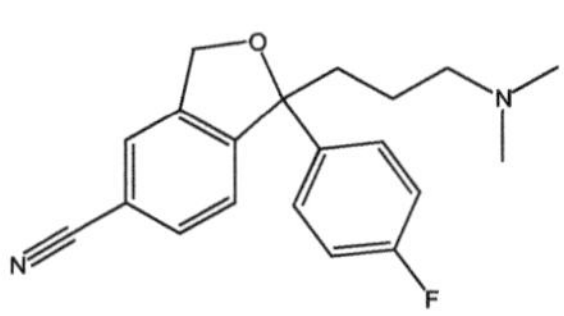	Not reported	2 TPs: *N*-desmethylcitalopram, citalopram-*N*-oxide [60]
Clinafloxacin (antimicrobial)	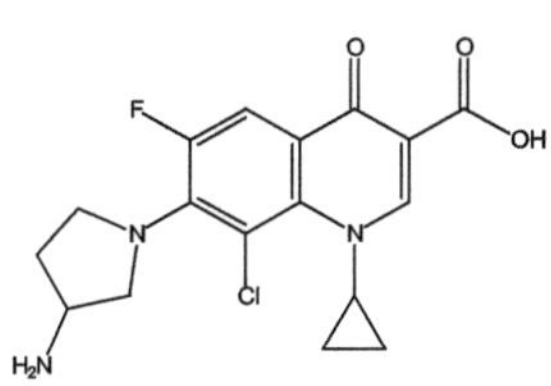	Not reported	8 TPs [16]
Clofibric acid (bLOOD lipid regulator)	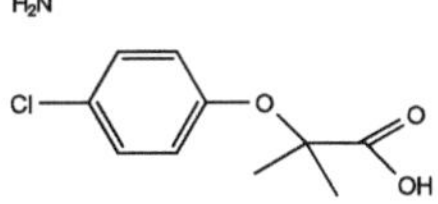	Not reported	4 TPs: hydroquinone, *b*-enzoquinone, phenol, 4-chlorophenol [61]

Table 2 (continued)

Pharmaceutical (therapeutic class)	Structure	Identified microbial degradates in the aquatic environment and in activated sludge	Identified photochemical degradates (environmental conditions)
Diatrizoate (X-ray contrast media)	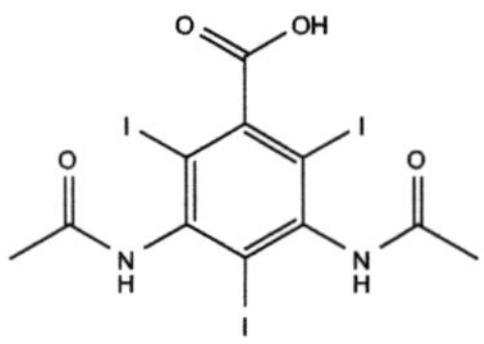	Activated sludge: 3-acetylamino-5-amino-2,4,6-triiodobenzoate. Water-sediment system: 3-acetylamino-5-amino-2,4,6-triiodobenzoate; 3,5-diamino-2,4,6-triiodobenzoate [62, 63]	Not reported
Diclofenac (anti-inflammatory)	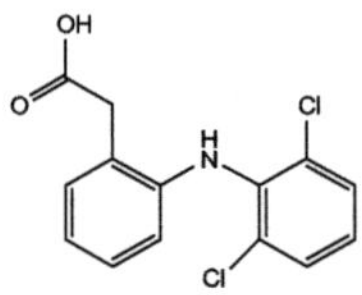	Not reported	3TPs [64] 11 TPs [65]
Enalapril (cardiovascular)	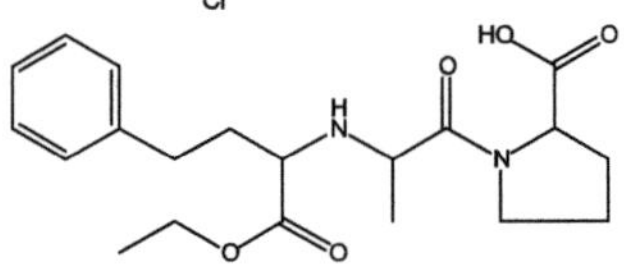	Not reported	3 TP [18]
Enalaprilat (cardiovascular, human metabolite of enalapril)	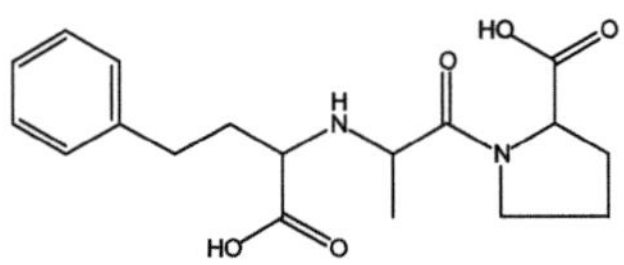	Not reported	1 TP [18]
Fluoxetine (psychiatric drug)	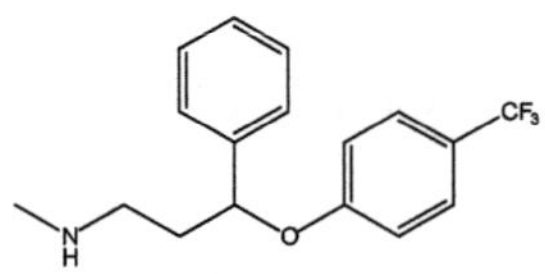	Not reported	4 TPs [20]
Furaltadone (antimicrobial)	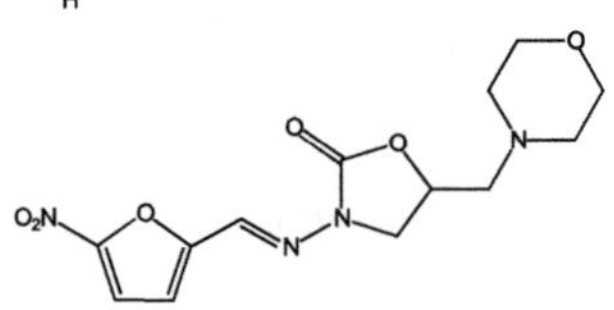	Not reported	3 TPs [66]
Furazolidone	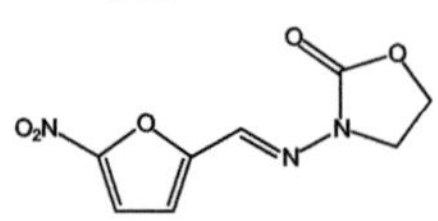	Not reported	3 TPs [66]

Table 2 (continued)

Pharmaceutical (therapeutic class)	Structure	Identified microbial degradates in the aquatic environment and in activated sludge	Identified photochemical degradates (environmental conditions)
Furosemide (diuretic)	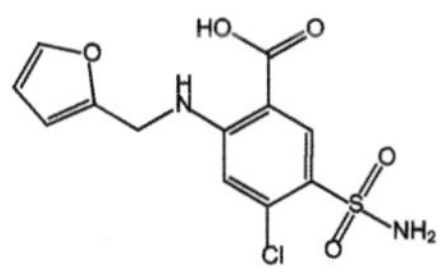	Not reported	1 TP [67]
Ibuprofen (analgesic)	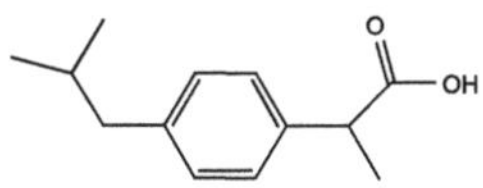	Activated sludge: hydroxyibuprofen, carboxy-ibuprofen, carboxy-hydratropic acid [68, 69]	Not reported
Iopromide (X-ray contrast media)	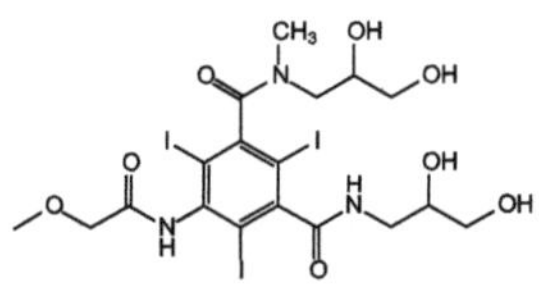	Activated sludge: carboxyiopromide; dicarboxylated iopromide; bis-dehydroxy-iopromide [21]; 5-amino-*N*,*N*′-bis(2,3-dihydroxy-propyl)-2,4,6-triiodo-*N*-methylisophthal-amide [70]	Not reported
Ketoprofen (analgesic)	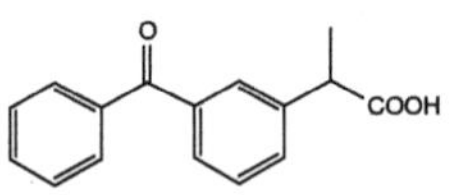	Activated sludge: 3-(hydroxyl-carboxy-methyl) hydratopic acid; 3-(keto-carboxy-methyl)-hydratopic acid [23]	3 TPs [71]
Levofloxacin (antimicrobial)	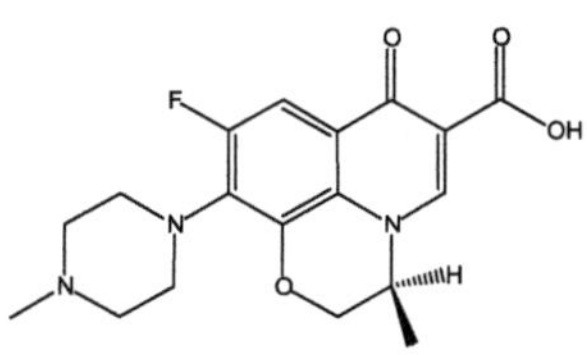	Not reported	3TPs [57]

Table 2 (continued)

Pharmaceutical (therapeutic class)	Structure	Identified microbial degradates in the aquatic environment and in activated sludge	Identified photochemical degradates (environmental conditions)
Mestranol 17α-Ethynyl estradiol (estrogen)	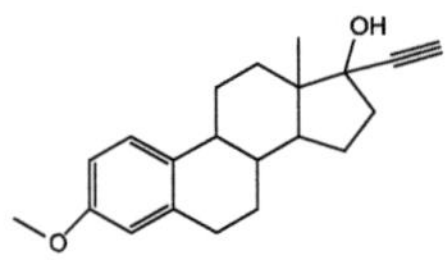 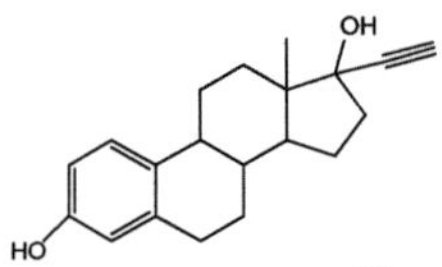 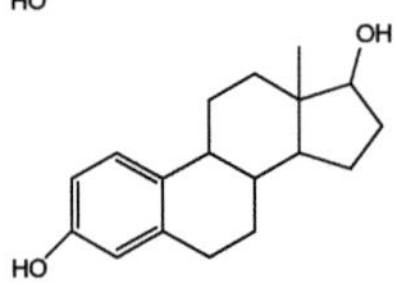	Activated sludge: mestranol → 17α-ethynylestradiol; 17β-estradiol → estrone [72]	Not reported
Methotrexate (anti-tumoral)	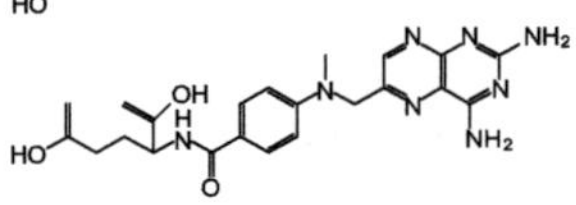	Activated sludge: 7-hydroxymethotrexate [73]	Not reported
Naproxen (analgesic)	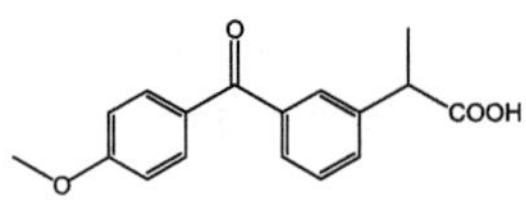	Activated sludge: *O*-desmethyl-naproxen [23]	TPs [74]
Nitrofurantoin (antimicrobial)	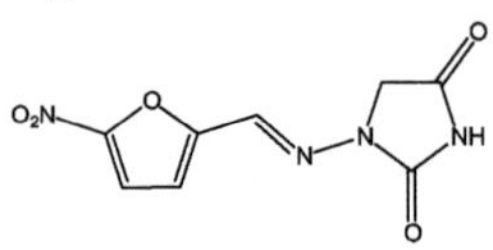	Not reported	3 TPs [66]
Propanolol (β-blocker)	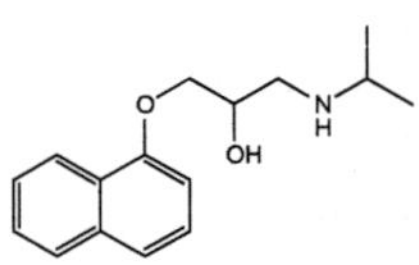	Not reported	3 TPs [75]
Sulfachloro-pyridazine	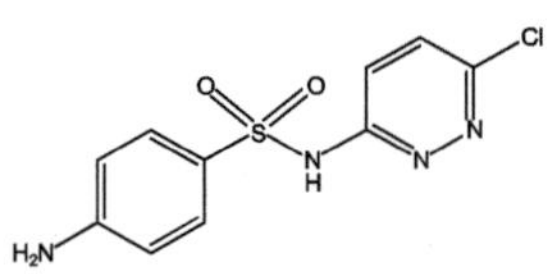	Not reported	2 TPs [76]
Sulfadiazine (antimicrobial)	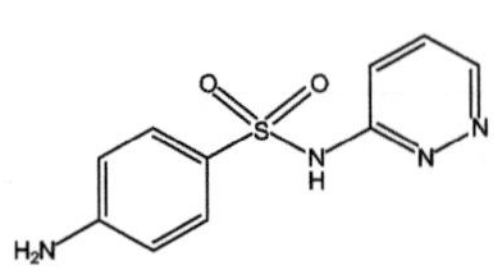	Not reported	1 TP [76]

Table 2 (continued)

Pharmaceutical (therapeutic class)	Structure	Identified microbial degradates in the aquatic environment and in activated sludge	Identified photochemical degradates (environmental conditions)
Sulfadimethoxine (antimicrobial)	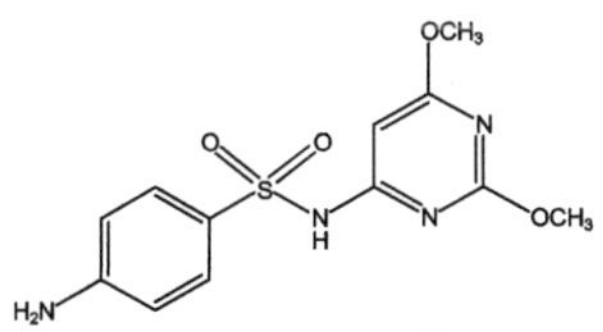	Not reported	1 TP [76]
Sulfamerazine (antimicrobial)	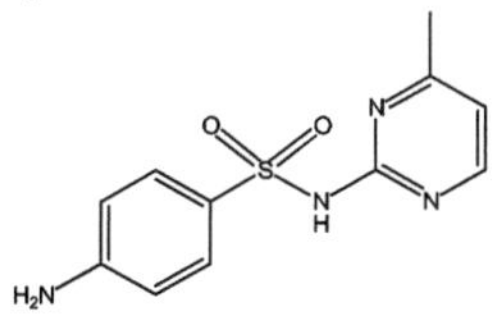	Not reported	1 TP [76]
Sulfamethazine (antimicrobial)	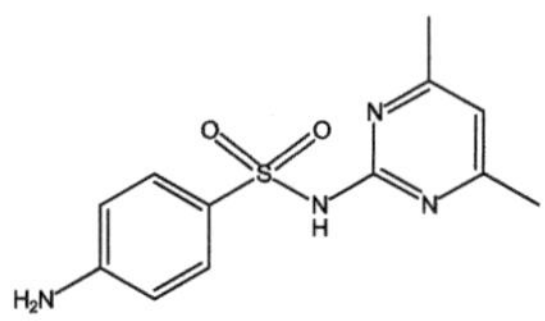	Not reported	1 TP [76]
Sulfamethizole (antimicrobial)	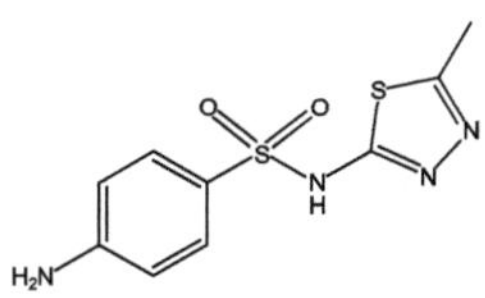	Not reported	2 TPs [77]
Sulfamethoxazole (antimicrobial)	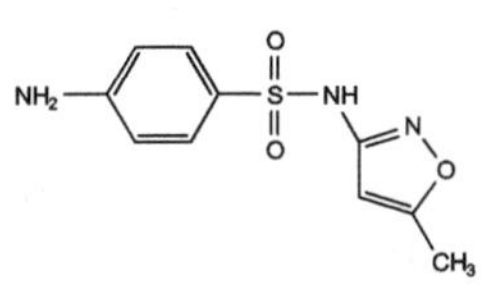	Not reported	2 TPs [77]
Sulfathiazole (antimicrobial)	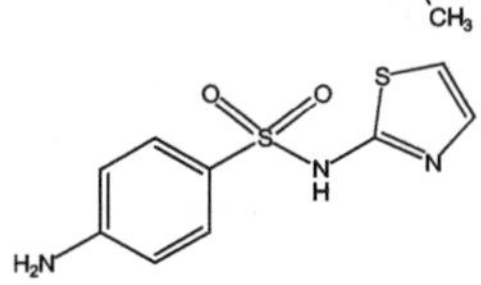	Not reported	2 TPs [77]
Sulfisoxazole (antimicrobial)	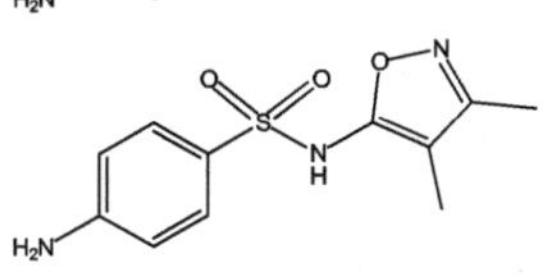	Not reported	2 TPs [77]

Table 2 (continued)

Pharmaceutical (therapeutic class)	Structure	Identified microbial degradates in the aquatic environment and in activated sludge	Identified photochemical degradates (environmental conditions)
Triclosan (antibacterial and preservative agent)	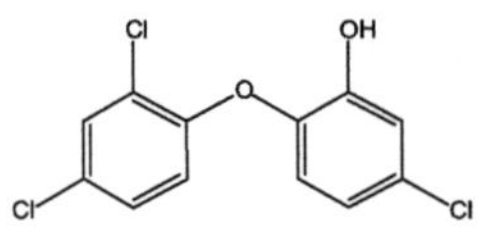	Activated sludge: methyltriclosan [78]	2 TPs [79] 2 TPs [80]
Trimethoprim (antimicrobial)	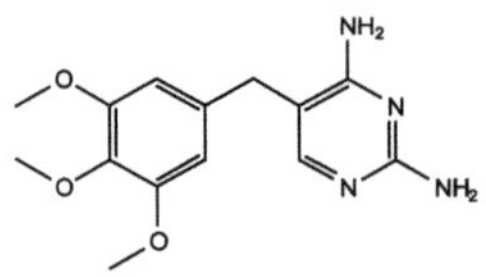	Activated sludge: α-hydroxy-trimethoprim; hydroxylated trimethoprim [81]	Not reported

taminate the soil though manure excretion or through the application of livestock slurries into the field. Three metabolites were identified by LC-FL and LC-QqQ-MS in APCI mode (see Table 3). The formation of the 7-amino danofloxacin derivative (TP-1) and 1-cyclopropyl-6-fluoro-7-amino-4-oxo-1,4-dihydroquinoline-3-carboxylic acid (TP-3) by cultures of *Candida lipopytica*, *Pseudomonas fluorescens*, two *Mycobacterium* species, and three *Penicillium* species demonstrates the propensities of these cultures to completely degrade the piperazine ring. Two *Mycobacterium* species, two *Pseudomonas* species, and isolates of *Nocardia* sp., *Rhizopus arrhizus* and *Streptomyces griseus*, all formed *N*-desmethyldanofloxacin (TP-2).

The combination of LC-QqLIT-MS and UPLC-QqToF-MS was applied to the structural characterization of the phototransformation products in aqueous media of the prodrug enalapril (376 Da, $C_{20}H_{28}N_2O_5$) and its active metabolite enalaprilat (348 Da, $C_{18}H_{24}N_2O_5$), a hypotensive cardioprotector (see Table 2) previously reported to occur in contaminated rivers [18]. The experiments yielded three detectable phototransformation products for enalapril (PTP-346 Da, 2 × 207 Da: PTP-207A and PTP-207B) whereas the photolysis of enalaprilat went hand in hand with the intermittent buildup of one phototransformation product, namely PTP-304 (see Table 3). Accurate mass measurements recorded on QqToF in MS/MS mode allowed the proposition of elemental compositions for the molecular ions of the transformation products (PTP-346 $C_{19}H_{26}N_2O_4$; PTP-207 $C_{12}H_{17}NO_2$; PTP-304 $C_{17}H_{24}N_2O_3$) as well as of their fragment ions.

To elucidate the structures of the four phototransformation products, the fragmentation patterns of both parent compounds were achieved. Upon

Table 3 Structures and characteristic fragment ions observed in MS techniques for identified transformation products of synthetic chemicals in the environment

Pharmaceutical (Class) [Refs.]	Atorvastatin (ATV) (Human pharmaceutical) [14]
Structure	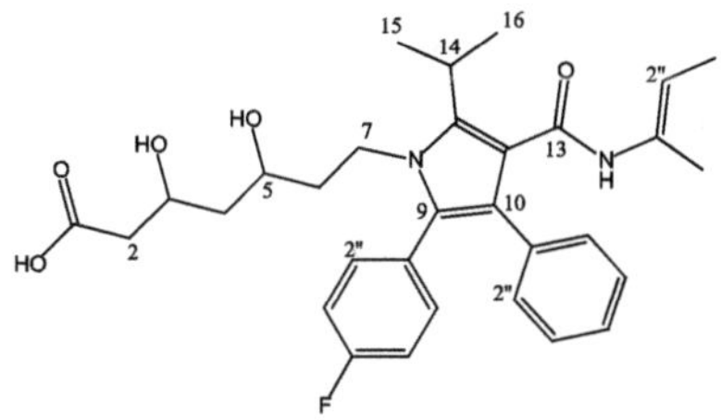
TP structure	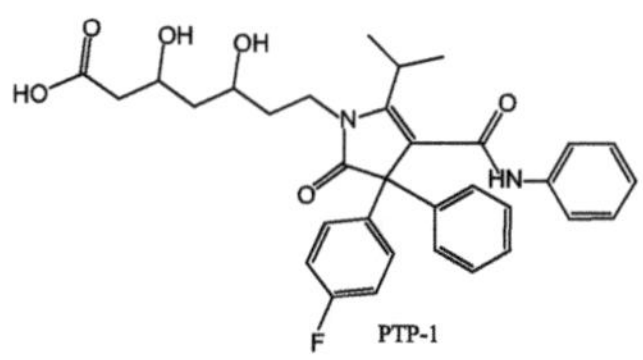 PTP-1 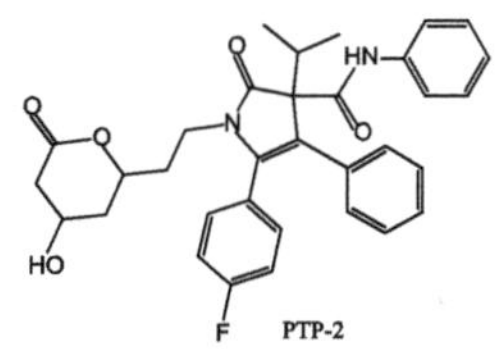 PTP-2 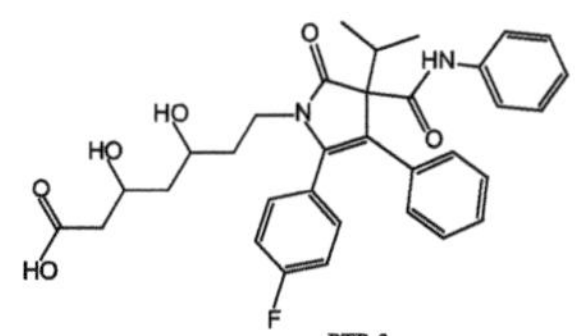 PTP-3 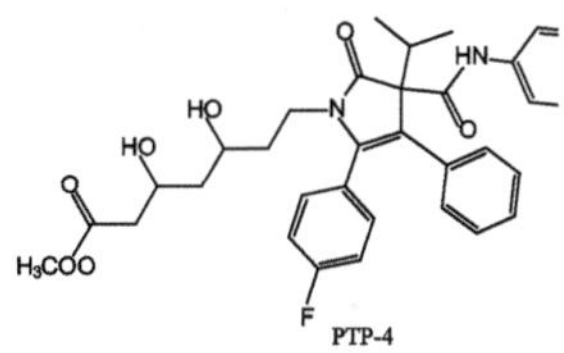 PTP-4
MS technique and product ions m/z	MALDI-ToF-MS ATV: $[M+H]^+$: m/z 581 PTP-1: $[M+H]^+$: m/z 575 PTP-2: $[M+Na]^+$: m/z 579 PTP-3: $[M+H]^+$: m/z 575 PTP-4: $[M+H]^+$: m/z 587
Transformation process	Direct photolysis
Type of reaction (Matrix)	Oxidation of the pyrrole ring and alkyl/aryl shift (Deionized water)

Table 3 (continued)

Pharmaceutical (Class) [Refs.]	Clinafloxacin (CLX) (Human pharmaceutical) [16]

Structure

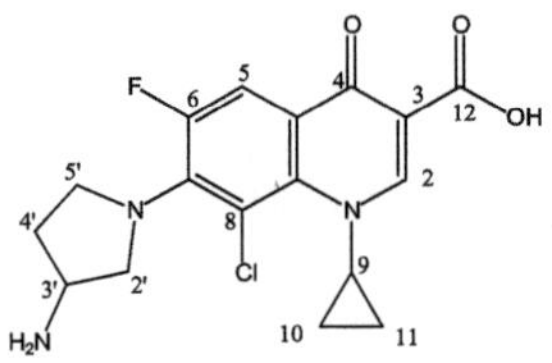

TP structure

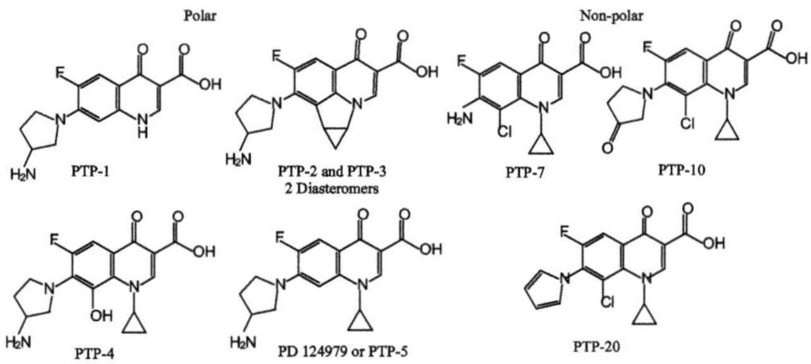

MS technique and product ions m/z

LC-MS, LC-IT-MS CLX	PTP-2
$[M+H]^+$: m/z 366	$[M+H]^+$: m/z 330
$[M+H-NH_3]^+$: m/z 349	$[M+H-NH_3]^+$: m/z 313
$[M+H-H_2O]^+$: m/z 348	$[M+H-H_2O]^+$: m/z 312
$[M+H-CO_2]^+$: m/z 322	$[M+H-CO_2]^+$: m/z 286
$[348-C_2H_5N]^+$: m/z 305	$[286-C_2H_5N]^+$: m/z 269

Transformation process	Direct photolysis;
Type of reaction	Dechlorination followed by further reactions on the quinolone ring, degradation of the pyrrolidine side-chain
(Matrix)	(Deionized water pH 4)

Table 3 (continued)

Pharmaceutical (Class) [Refs.]	Danofloxacin (DNX) (Veterinary drug) [17]
Structure	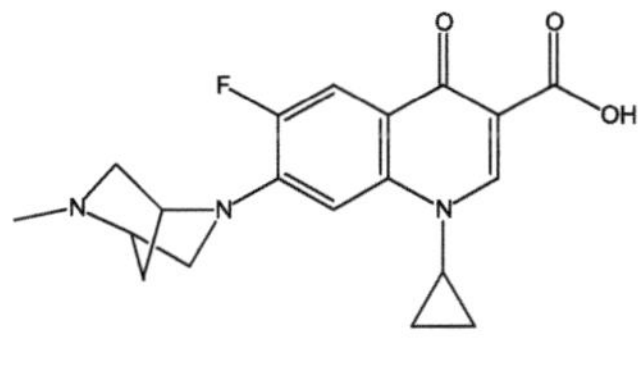
TP structure	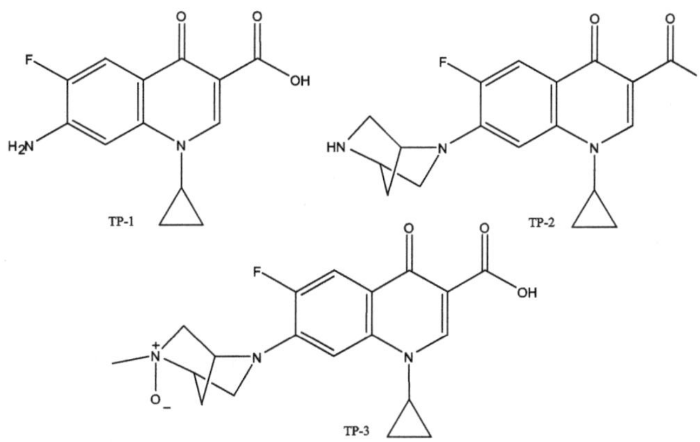
MS technique and product ions m/z	LC-QqQ-MS TP-1: $[M+H]^+$: m/z 263 TP-2: $[M+H]^+$: m/z 344
Transformation process	Microbial degradation
Type of reaction (Matrix)	*N*-Demethylation, cleavage at the piperazine ring and *N*-oxidation (Soil)

Table 3 (continued)

Pharmaceutical (Class) [Refs.]	Enalapril (ENAL) (Human pharmaceutical) [18]		Enalaprilat (ENALAT) (TP of ENAL)	
Structure	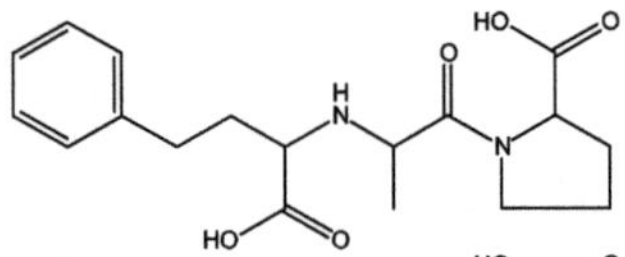 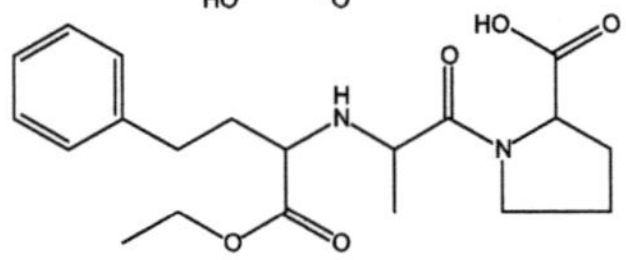			
TP structure	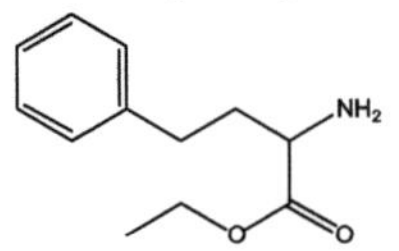 PTP-207A		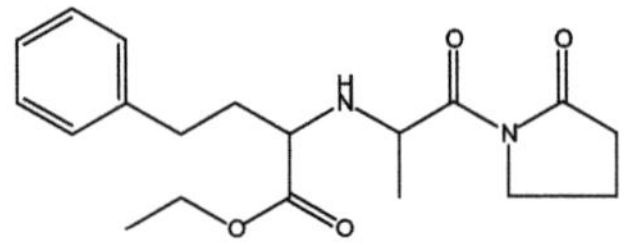 PTP-346	
	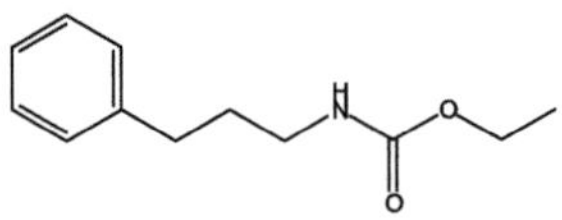 PTP-207B		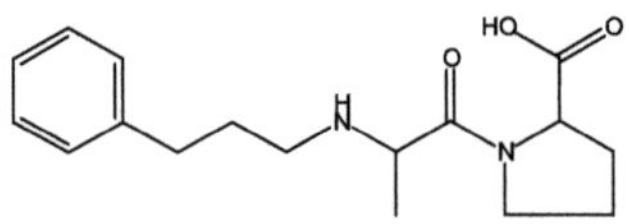 PTP-304	
MS technique and product ions m/z	LC-QqLIT-MS and UPLC-QqToF-MS			
	ENAL	ENALAT	PTP-207A	PTP-207B
	$[M+H]^+$: m/z 377	$[M+H]^+$: m/z 349	$[M+H]^+$: m/z 208	$[M+H]^+$: m/z 208
	$[M+H-(HCOOC_2H_5)]^+$: m/z 303	$[M+H-(HCOOH)]^+$: m/z 303		
	-	-	$[M+H-(NH_3)]^+$: m/z 191	-
	$[M+H-(C_8H_{13}NO_3)]^+$: m/z 206	$[M+H-(C_8H_{12}NO_3)]^+$: m/z 178	$[M+H-(CH_2CH_2)]^+$: m/z 180	-
	$[M+H-(C_{11}H_{17}NO_5)]^+$: m/z 134	$[M+H-(C_9H_{12}NO_5)]^+$: m/z 134	$[M+H-(CH_3-CH_2-O-CHO)]^+$: m/z 134	$[M+H-(CH_3-CH_2-O-CHO)]^+$: m/z 134
Transformation process	Direct photolysis			
Type of reaction (Matrix)	Cleavage at the secondary amine, conversion of the acid into ketone (Synthetic surface water)			

Table 3 (continued)

Pharmaceutical (Class) [Refs.]	Fipronil (Insecticide) [19]	Desulfinyl-fipronil (PTP)

Structure

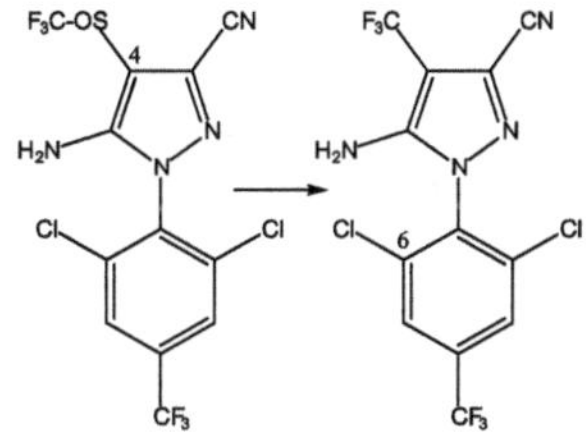

TP structure

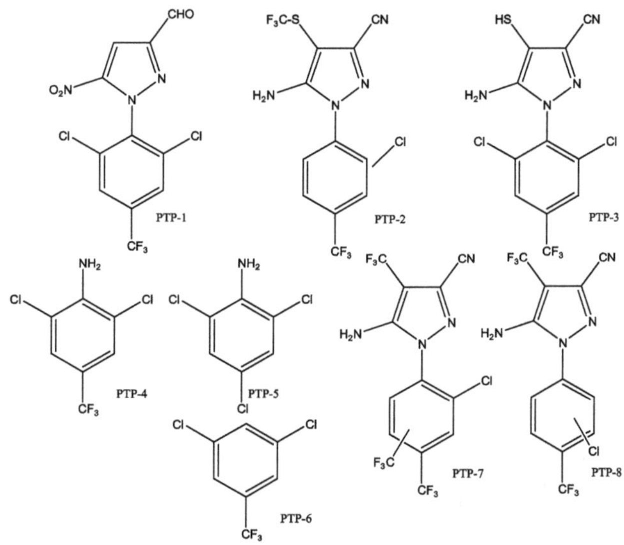

MS technique and product ions m/z

GC-EI-MS after silylation

PTP-1	PTP-2	PTP-3
$[M-1]^+$: m/z 353	$[M]^+$: m/z 386	$[M]^+$: m/z 352
$[C_6H_2(Cl_2)(CF_3)(N_2C_3H)(NO_2)]^+$: m/z 325	$[M-F_2(NH_2)]^+$: m/z 333	$[M-C_2N(SH)(CN)]^+$: m/z 255
$[C_6H_2(Cl_2)(CF_3)(N_2C_3H]^+$: m/z 279	$[C_6H_3(CF_3)ClC_3N_2(NH_2)(CN)S]^+$: m/z 317	$[C_6H_2(Cl_2)(CF_3)]^+$: m/z 213

Transformation process	Direct photolysis
Type of reaction	Cleavage of the pyrazole ring, desulfuration and rearrangements in the aromatic ring
(Matrix)	(Water/ethanol)

Table 3 (continued)

Pharmaceutical (Class) [Refs.]	Fluoxetine (Human pharmaceutical) [20]		
Structure	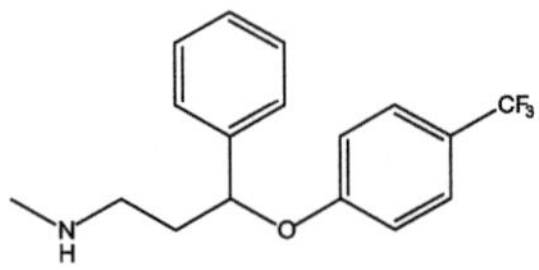		
TP structure	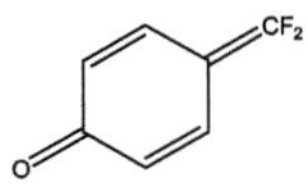 PTP-1 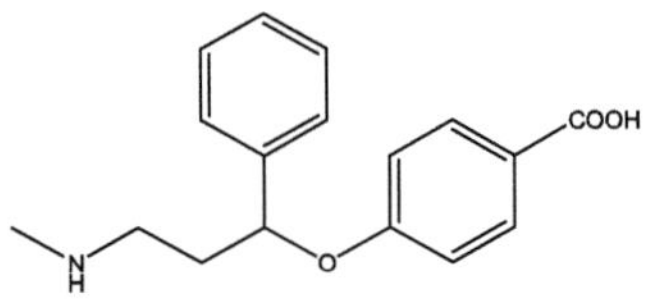 PTP-3 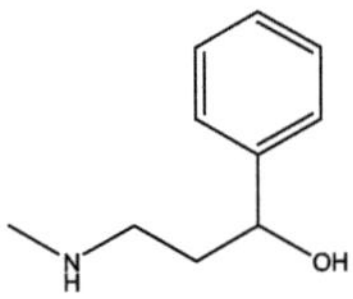 PTP-2 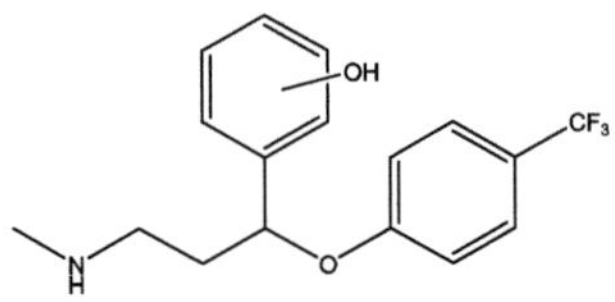 PTP-4		
MS technique and product ions m/z	LC-QqQ-MS PTP-2	PTP-3	PTP-4
	$[M+H]^+$: m/z 166 $[M+H-H_2O]^+$: m/z 148	$[M+H]^+$: m/z 286 $[M+H-H_2O]^+$: m/z 268	$[M+H]^+$: m/z 326 $[M+H-H_2O]^+$: m/z 308
Transformation process	Direct and indirect photolysis		
Type of reaction (Matrix)	O-Dealkylation, defluorination and $^{\cdot}$OH adition (Deionized water and synthetic surface water)		

Table 3 (continued)

Pharmaceutical (Class) [Refs.]	Iopromide (Human pharmaceutical) [21]			
Structure	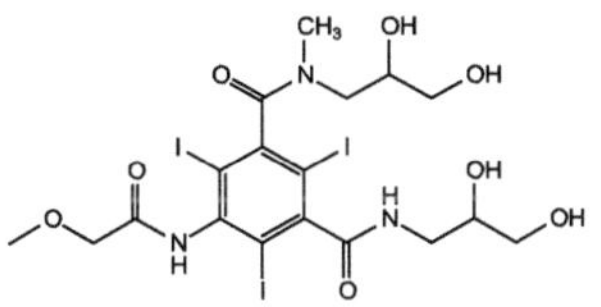			
TP structure	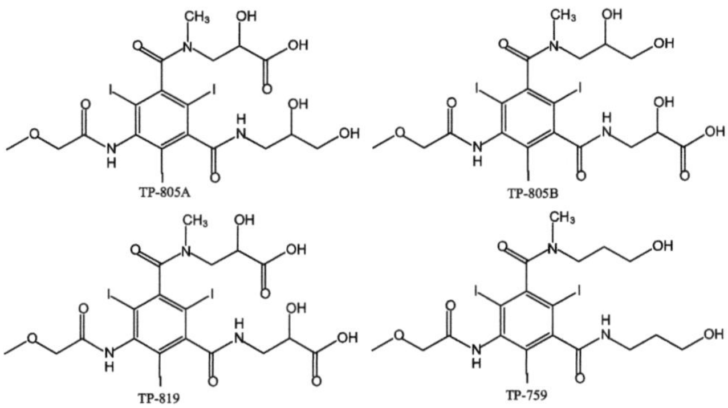			
MS technique and product ions m/z	LC-IT-MS including H/D exchange			
	IOP	TP-805A	TP-805B	TP-759
	$[M+H]^+$: m/z 792	$[M+H]^+$: m/z 806	$[M+H]^+$: m/z 806	$[M+H]^+$: m/z 760
	$[M+H-H_2O]^+$: m/z 774	$[M+H-H_2O]^+$: m/z 788	$[M+H-H_2O]^+$: m/z 788	-
	$[M-CH_3-NH-CH_2-CH-OH-CH_2-OH]^+$: m/z 687	$[M-CH_3-NH-CH_2-CH-OH-COOH]^+$: m/z 687	-	$[M-CH_3-NH-CH_2-CH-CH_2-OH]^+$: m/z 671
	$[687-HI]^+$: m/z 559	$[687-HI]^+$: m/z 559	-	$[671-HI]^+$: m/z 543
Transformation process	Microbial degradation			
Type of reaction (Matrix)	Oxidation and dehydroxylation (Wastewater)			

Table 3 (continued)

Pharmaceutical (Class) [Refs.]	Ketoprofen (KTF) (Human pharmaceutical) [23]
Structure	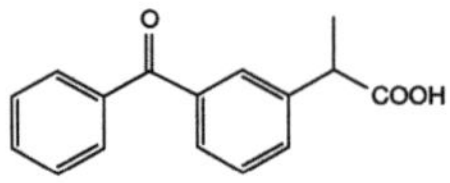
TP structure	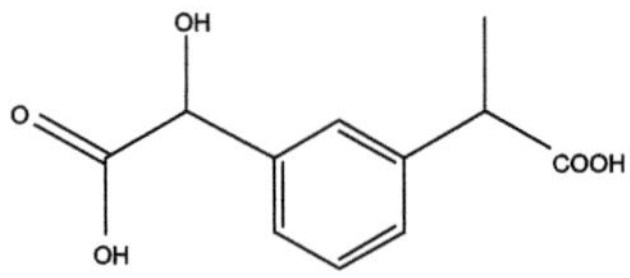 TP-1 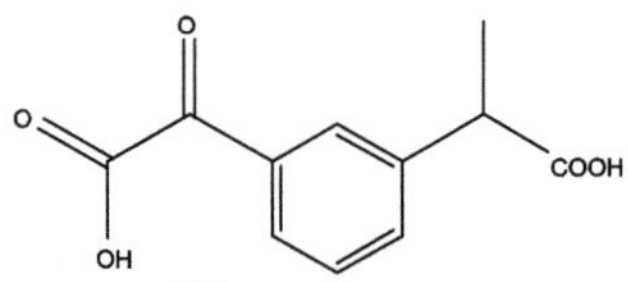 TP-2
MS technique and product ions *m*/*z*	LC-QqQ-MS

KTF	TP-1	TP-2
$[M\text{-}H]^-$: *m*/*z* 253	$[M\text{-}H]^-$: *m*/*z* 223	$[M\text{-}H]^-$: *m*/*z* 221
$[M\text{-}H\text{-}CO_2]^-$: *m*/*z* 209	$[M\text{-}H\text{-}CO_2]^-$: *m*/*z* 179	$[M\text{-}H\text{-}CO_2]^-$: *m*/*z* 177
–	$[179\text{-}CO_2]^-$: *m*/*z* 135	$[177\text{-}CO_2]^-$: *m*/*z* 133

Transformation process	Microbial degradation
Type of reaction (Matrix)	Oxidative ring opening (Wastewater)

comparing the (+)-ESI-MS[2] spectra of enalapril and enalaprilat recorded in enhanced product ion (EPI) mode, two series of fragment ions could be distinguished: one series comprised ions that were common to both compounds, namely *m*/*z* 303, 160, 134, and 117 (see Table 3) while in the other one, ion pairs were identified that differed by *m*/*z* 28 (*m*/*z*234/206, *m*/*z* 206/178,

Table 3 (continued)

Pharmaceutical (Class) [Refs.]	Naproxen (NPX) (Human pharmaceutical) [23]	
Structure	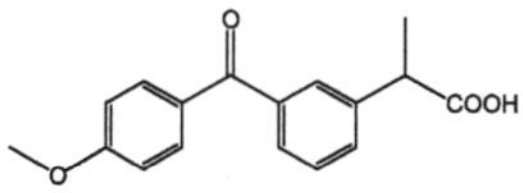	
TP structure	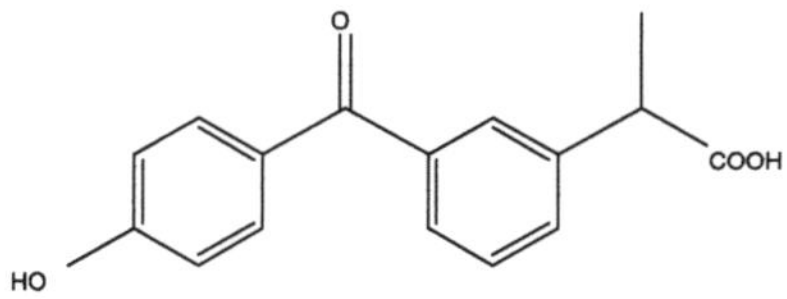 TP-1	
MS technique and product ions m/z	LC-QqQ-MS NPX	TP-1
	$[M-H]^-$: m/z 230 $[M-H-CO_2]^-$: m/z 286	$[M-H]^-$: m/z 215 $[M-H-CO_2]^-$: m/z 171
Transformation process	Microbial degradation	
Type of reaction (Matrix)	O-Demethylation (Wastewater)	

and m/z 130/102), i.e., fragment ions containing the ester (R=CH_3–CH_2) or the acid functionality (R=H). Regarding the spectra of the two major photodegradation products of enalapril (PTP-207A and PTP-207B), fragment ions with nominal masses m/z 117 and 134 were also observed in the QqToF-MS^2 spectrum of enalapril. This indicated that the left part of the molecule comprising the aromatic ring was intact in both transformation products. Based on the proposed elemental composition of the protonated PTP-207A, $C_{12}H_{18}NO_2$ (– 4.3 ppm), a plausible structure corresponded to the cleavage between the nitrogen of the secondary amine and the α-carbon of the amide group [18]. The presence of a primary amine in PTP-207A facilitated the neutral loss of ammonia, resulting in m/z 191 ($C_{12}H_{15}O_2$, – 2.0 ppm) while the ion m/z 180 ($C_{10}H_{14}NO_2$, + 1.5 ppm) was identified as corresponding to the elimination of ethylene in the ester side chain, accomplished through a McLafferty rearrangement within this moiety; the loss of both ammonia and ethylene resulted in the formation of m/z 163 ($C_{10}H_{11}O_2$, + 1.5 ppm).

Table 3 (continued)

Pharmaceutical (Class) [Refs.]	Sulfadiazine (SDZ) (Antimicrobial) [25]		
Structure	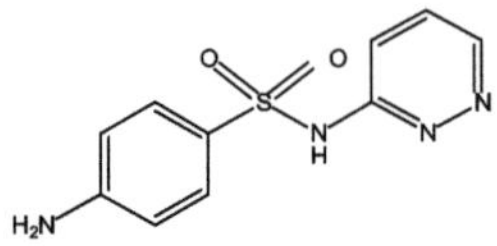		
TP structure	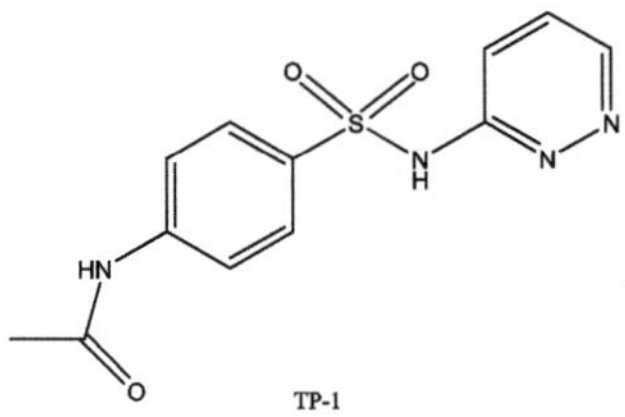 TP-1	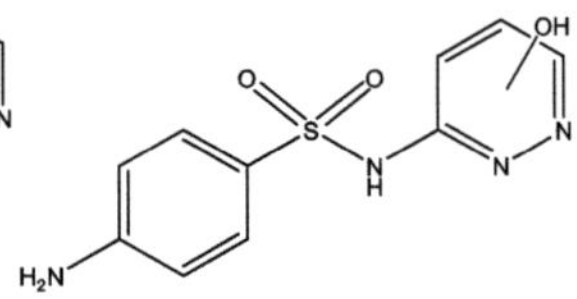 TP-2	
MS technique and product ions m/z	LC-QqQ-MS SDZ	TP-1	TP-2
	$[M+H]^+$: m/z 251	$[M+H]^+$: m/z 293	$[M+H]^+$: m/z 268
	$[M+H-(H_2NC_4H_3N_2)]^+$: m/z 156	$[M+H-(H_2NC_4H_3N_2)]^+$: m/z : 198	$[M+H-(H_2NC_4H_3N_2)]^+$: m/z 156
Transformation process	Microbial degradation		
Type of reaction (Matrix)	N-Acetylation and hydroxylation (Soil)		

With respect to the phototransformation product PTP-207B, accurate mass measurements on the QqToF instrument suggested that its elemental composition was identical to that of PTP-207A ($C_{12}H_{18}NO_2$ for the protonated compound, – 4.3 ppm), i.e., it had to be a constitutional isomer. The absence of a fragment ion at m/z 191 in the MS^2 spectrum of PTP-207B indicated that the neutral loss of ammonia from the protonated precursor ion was not feasible. Thus, the presence of a primary amine in this molecule could be ruled out. On the other hand, the compound had to include a structural element that could be easily cleaved off to yield the fragment ion m/z 134 [18]. This ion had been detected and identified previously in the MS^2 spectrum of both enalapril and enalaprilat. A migration of the ester side chain in PTP-207A to the nitrogen atom was put forward to explain the formation of the carbamate structure of the transformation product PTP-207B. As the weakest bond in this molecule was thought to be the nitrogen–carbon bond of the carbamate, the formation of m/z 134 was favored over other fragmentation routes. This resulted in a rather simple fragmentation pattern of PTP-207B in the positive ion mode. The other two transformation products were also identified with the comparison of MS spectral data (data not shown). As to the analytical approach employed, the use of LC-QqLIT-MS and UPLC-QqToF-MS proved to be a powerful combination for elucidating the unknown structures of the phototransformation products [18].

Fipronil is a recently discovered insecticide of the phenylpyrazole series and desulfinyl-fipronil is its major PTP [19]. In this study, the photodegradation of fipronil and desulfinyl-fipronil was studied with GC-EI-MS previous silylation, at low light intensities (sunlight or UV lamp). In the photodegradation experiments of fipronil, six new PTPs were identified and another four had been previously described, based on the similarities in their chemical structures. Fipronil photodegradation occurs via at least two distinct pathways, one of which involves desulfuration at the 4-position of the pyrazole ring giving the desulfinyl derivative (PTP-1). The other pathway involves a different modification of the 4-substituent (PTP-2–PTP-6), leading to cleavage of the pyrazole ring and formation of the aniline derivatives as well as to the formation of sulfide derivatives(see Table 3) [19]. The PTP-1 with a mass of m/z 354 was unable to be silylated. The fragmentation pattern presented the loss of H from the molecular ion ($[M\text{-}1]^+$: m/z 353), the loss of an aldehyde group ($[C_6H_2(Cl_2)(CF_3)(N_2C_3H)(NO_2)]^+$: m/z 325), and the loss of NO_2 ($[C_6H_2(Cl_2)(CF_3)(N_2C_3H)]^+$: m/z 279). Two novel sulfide derivatives were also detected (PTP-2 and PTP-3).

A minor product that can be monosilylated and showed a fragmentation pattern similar to that of desulfonyl-fipronil corresponded to the sulfide-fipronil (PTP-2) with a monochlorinated phenyl group (m/z 386, 255 and 213). Another PTP was detected (PTP-3), which presented a mass of 353 and which was able to be disilylated. The structure proposed on the basis of MS analysis (m/z 352, 255, and 213) was 4-thiol-fipronil, a sulfonic acid derivative (see Table 3). Finally, among the studied derivatives in the reaction mixture, compounds lacking the pyrazole ring (see Table 3) were detected (MS data not shown). These corresponded to aniline derivatives (PTP-4–PTP-6) with various chlorine substitutions on the benzene ring, thus testifying to the high degree of fipronil degradation under UV irradiation. Exposure of hydro-alcoholic solution of fipronil to UV light or sunlight led to the formation of desulfinyl-fipronil as the main product, as reported previously. When pure desulfinyl-fipronil was exposed to UV light, the formation of two other derivatives (PTP-7 and PTP-8) was observed, resulting from rearrangements on the aromatic ring (see Table 3). The PTP-8 had lost a chlorine substituent from the aromatic ring, whereas, in the PTP-7, the chlorine atom at the aromatic position 6 had been substituted by a CF_3 group. No cleavage of the pyrazole ring was observed. Tests comparing the effects of UV lamp irradiation to solar irradiation allowed the authors to conclude that the in vitro conditions chosen were qualitatively representative of what may occur in nature [19].

Fluoxetine, a psychiatric drug that is one of the most frequently prescribed pharmaceuticals of this class, was demonstrated to be susceptible to photodegradation in surface waters [20]. LC-QqQ-MS was used for the characterization of four degradation products. Direct photodegradation of fluoxetine (see Table 2) led to the generation of two *O*-dealkylated products (PTP-1 and PTP-2) and one carboxylic acid product (PTP-3) formed by defluorination of the trifluoromethyl group in the parent compound (see Table 3). The retention times and the MS spectra of the O-dealkylated products were in agreement with those obtained for the authentic standards. In the ESI-MS^2 spectrum of PTP-2, a molecular ion was observed at m/z 166 and two product ions at m/z 44, which corresponds to the formation of $[H_2C{=}NH{-}CH_3]^+$, and m/z 148, which corresponds to the loss of water. The third compound formed under direct photolysis was also analyzed in ESI(+) mode and the molecular ion at m/z 286 as well as a product ion corresponding to the loss of water (m/z 268) were observed. Under hydroxyl radical-producing conditions (indirect photodegradation), a new product was formed resulting from ring addition of $^{\bullet}OH$ to the parent compound. The molecular ion observed in ESI positive mode at m/z 326 is 16 amu greater than the fluoxetine (m/z 310) suggesting that an oxygenated photoproduct was formed. The product ion scan revealed the characteristic ion at m/z 44, corresponding to the alkylamine side chain and a M-18 ion at m/z 308 $[M{-}H_2O]^+$, which is a characteristic loss for compounds containing alcohol functional groups. Furthermore, the m/z 164 product ion corresponds to $HO{-}C_6H_4{-}CH{-}(CH_2)_2NHCH_3$ [20].

The combination of LC-MS and LC-IT-MS and H/D exchange [21] was used for the study of the biotransformation of iopromide in conventional activated sludge and nitrifying activated sludge collected from a municipal wastewater treatment plant (WWTP). Four metabolites were identified. In a batch-reactor containing mixed liquor from a conventional activated sludge, three metabolites produced upon oxidation of the primary alcohols to carboxylic acids in the side chains of iopromide (m/z 791) were identified. Two metabolites (m/z 806) carried one carboxy group in either side chain, while the third one had one carboxy groups in each side chain (m/z 820). Furthermore, in a batch-reactor with mixed liquor from the nitrification tank of a WWTP, one metabolite (m/z 760) was identified as a dehydroxylation product. As to the fragmentation pattern of iopromide, selection of the protonated molecule (m/z 792) showed five main fragment ions m/z 774 (loss of water), m/z 701 and m/z 687 (cleavage of the amide bond in different side chains), and m/z 573 and m/z 559 formed out of the latter two by loss of HI. Key fragment ions for the identification of the two metabolites at m/z 806 (14 Da higher than iopromide) and at m/z 820 (28 Da higher than iopromide) were the loss of water (*m*/*z* 788 and *m*/*z* 802) and the neutral loss of one HCOOH (46 Da) for *m*/*z* 806 and two HCOOH molecules for the metabolite with *m*/*z* 820. The detection of the fragment ions *m*/*z* 687 and 559 in the MS^2 spectrum of TP-805A (Table 3) indicated that the side chain B was still intact because these two ions were also observed in the mass spectrum of iopromide. On the other hand, the ion masses *m*/*z* 701 and 573 observed in the MS^2 spectrum of iopromide increased by 14 units, resulting in *m*/*z* 715 and 587 in TP-805A. This suggested that a chemical modification in the side chain A occurred during metabolism. Conversely, the disappearance of the fragment ions *m*/*z* 687 and 559 in the MS^2 spectrum of TP-805B (Table 3) as compared to iopromide could be rationalized as an increase of 14 mass units in the side chain B. The mass spectrum of the metabolite at *m*/*z* 760 (32 Da lighter than iopromide) showed the formation of *m*/*z* 685 and 671, attributed to breakage of the amide bonds in the side chains, and *m*/*z* 557 and 543, corresponding to further losses of HI. Compared to the MS^2 spectrum of iopromide, the mass shift by – 16 Da in the fragment ions *m*/*z* 685 and 671 indicated that both side chains had undergone structural changes during biotransformation [21]. The absence of a neutral loss of water from $[M+H]^+$ in the MS^2 mass spectrum of *m*/*z* 760 (in contrast to the MS^2 spectra of iopromide and the metabolites *m*/*z* 806 and *m*/*z* 820) was a strong indicator that dehydroxylation of the secondary hydroxy group in the side chain occurred. H/D-exchange experiments and derivatization of the carboxy groups to methyl esters provided definitive confirmation for the metabolite identities. Overall, the iodinated ring of iopromide remained intact during the biodegradation processes, in contrast to the outcomes reported in [22].

For the identification of the degradation products of two analgesics, ketoprofen and naproxen, in activated sludge from a municipal WWTP, LC-DAD and LC-(–)ESI-MS/MS were used [23]. For co-metabolic transformation, milk powder was added to the test medium, while in a second experimental design the pharmaceutical compound served as a sole carbon source. Of the two pharmaceuticals, ketoprofen was the only compound biotransformed in the reactor without milk powder. During ketoprofen degradation, there were two novel metabolites 3-(hydroxy-carboxymethyl) hydratropic acid (TP-1) and its oxidative form 3-(keto-carboxymethyl)-hydratropic acid (TP-2) (Table 3). TP-1 showed an UV spectrum with only one absorption band around 220 nm. The (–)MS/MS spectrum was characterized by a deprotonated molecule at *m*/*z* 223, which initially lost two CO_2 molecules to form *m*/*z* 179 and *m*/*z* 135, followed by the loss of H_2O to form *m*/*z*117 (Table 3). These fragmentations were suggestive of the presence of two carboxy groups and one aliphatic hydroxy group in the metabolite. The UV spectrum of TP-2 was characterized by two absorption bands, indicating higher conjugation than TP-1 [23]. Additionally, the MS/MS fragmentation from the precursor ion (*m*/*z* 221) was characterized by the loss of two CO_2 molecules (*m*/*z* 177 and *m*/*z* 133) and subsequently CO (*m*/*z* 105). Based on these metabolites, it was proposed that ketoprofen was degraded according to the pathway known for biphenyls, biphenyl ethers, and related compounds. TP-2 remained stable throughout the experimental period of 28 days. In the experiment conducted with an additional carbon source to identify co-metabolic degradation, ketoprofen was not degraded. As for the degradation of naproxen, about 60% of biotransformation occurred within 28 days of the experiment producing one detectable metabolite (TP-1) [23]. It was identified as *O*-desmethyl-naproxen based on its deprotonated molecule (m/z 215) and the product ion spectrum, which indicated the presence of one carboxy group. This methyl ether cleavage is a well-documented metabolic pathway in mammals [24]. This metabolite was unstable under the experimental conditions.

The transformation of sulfadiazine in soil using ^{14}C-labeled sulfadiazide was reported [25]. Among other veterinary pharmaceuticals, sulfadiazine is widely used as an antimicrobial substance in intensive livestock production to prevent and treat diseases. The authors also detected up to three transformation products in the batch-type studies. Two of these were identified as acetyl- and hydroxy-sulfadiazine (TP-1 and TP-2), respectively; the other remained unidentified. TP-1 has the acetyl group attached to the amino group connected to the phenyl ring and in TP-2 the hydroxyl group is attached to the diazine ring. During metabolism in treated pigs, sulfadiazine is transformed to *N*-4-acetyl-sulfadiazine and both substances are consequently found in pig manure [26]. The alkylation of sulfadiazine is reversible in stored manure. Although these studies focused on the metabolism in manure this study indicated that transformation reactions might also occur in soils [26].

2.2 Quantitative Analysis

An extensive body of literature has been published with respect to the quantitative analysis of TPs of pesticides in air [27, 28], surface waters [29, 30], groundwater [31, 32], sediment [33, 34], and soil [35, 36] and several critical reviews on existing methods have been recently published [37, 38]. A number of multiresidue methods have been developed for the determination of parent compounds and TPs covering a wide range of pesticide classes [37]. Compounds that are typically included into monitoring programmes are organochlorines DDT and its TPs DDE and DDD; endosulfan, endrin, and heptachlor and their TPs; triazine herbicide atrazine and its TPs deethylatrazine and deisopropylatrazine; chloroacetanilide herbicides and their acidic TPs.

On the other hand, quantitative analyses of TPs of pharmaceuticals and related compounds in the environment have been rarely reported. In recent years, pharmaceuticals have been the focus of global environmental research, which has resulted in a large number of analytical methods being developed for the quantitative analysis of pharmaceutical residues in surface

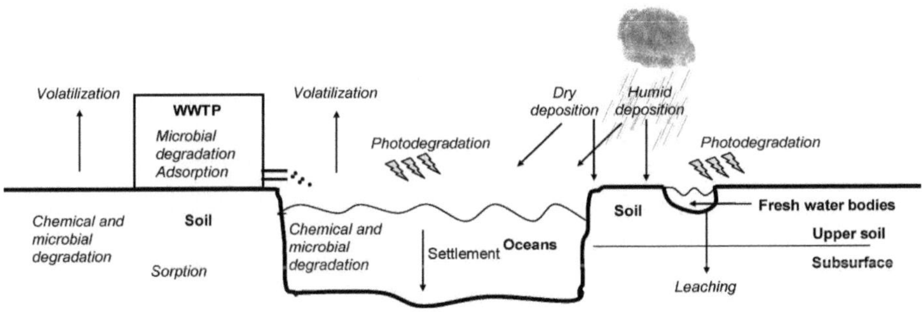

Fig. 1 Fate of organic pollutants in the environment

and wastewaters, as well as in soil and sediment (Fig. 1 shows the fate of organic pollutants in the environment). A relatively large number of review papers have been produced covering analytical issues related to their determination in environmental samples [39–42].

Most of the literature reviewed on the quantitative analytical methods used to determine pharmaceuticals in the environment have focused either on one class of compounds (e.g., β-blockers, antiinflammatory drugs, antibiotics) or, in the case of multiresidue methods, includes at the maximum 30–40 target analytes, which are in most cases parent compounds or their human metabolites. TPs are rarely included in monitoring programs and the simultaneous analysis of TPs has been reported in only a few studies. The main reasons for this are the lack of analytical methods, unavailability of standards, and the lack of data regarding their ecotoxic effects that will help in the prioritization of the analytical efforts. Therefore, a number of potentially relevant TPs are waiting to be brought under scrutiny.

2.2.1 Sample Extraction and Clean-Up

Due to the dilution and possible degradation of synthetic chemicals in the environment, low levels can be expected. Therefore, an analyte preconcentration procedure is almost always necessary in order to achieve desired levels of analytical sensitivity, often requiring high enrichment factors of between 100 and 10000. Such enrichment factors for analysis are usually achieved using solid-phase extraction (SPE). Different approaches can be followed in the SPE procedure depending on the target compounds: (i) the determination of a wide range of parent compounds and selected TPs (i.e., multiresidue methods for pesticides and their TPs), or (ii) a limited number of compounds usually belonging to the same class (i.e., methods for determination of chloroacetanilides and their corresponding ethanesulfonic acid (ESA) and oxanillic acid (OA) transformation products, or methods for oxytetracyclines and their TPs).

Table 4 Comparison of extraction techniques

Soxhlet	Sonication	PLE	MAE	MASE
Advantages				
• Easy to handle • Inexpensive equipment • No filtration required • High matrix capacity	• Inexpensive equipment • Easy to handle and to optimize • Fast extraction • Moderate solvent consumption (20–100 ml)	• Simple optimization procedure (easy to transfer an existing Soxhlet or sonication method) • Fast extraction • Low solvent consumption (20–30 ml) • No filtration required • High level of automation • Easy to use	• Several extractions can be performed simultaneously (up to 12) • High level of automation • Fast extraction (ca. 10–30 min) • Low solvent consumption (10–70 ml) • Moderate investment	• Fast extraction (10–60 min) • Low to moderate solvent volume (10–150 ml)
Drawbacks				
• Demands large volumes of highly purified solvents (up to 200 mL) • High costs of purchase and disposal of solvents • Long extraction times (up to 48 h) • generates dirty extracts • not automatable	• Filtration or/and centrifugation required • Depending on matrix needs repeated extractions to achieve good recovery • Not automatable	• High investment cost of the commercialized systems • Elevated temperatures may degrade thermolabile analytes	• Requires further filtration of the extracts • The solvent must absorb microwaves (unless water is present in the matrix) • Inhomogeneity of the field inside the cavity	• Extraction solvent must be able to absorb microwaves • Clean up required • Degradation of some pesticides

In the first approach, typically broad spectrum alkyl-silica or polymeric-based solid phases together with a cation-exchange solid phase, mixed mode or multilayer cation exchange/alkyl-based solid phase are used. Oasis HLB cartridges are preferred, working generally at neutral pH. Due to their chemical composition (the combination of the lipophilic divinylbenzene and the hydrophilic *N*-vinylpyrrolidone polymers), they are able to extract acidic, neutral, and basic compounds over a wide pH range, including neutral pH. C_{18} is another SPE sorbent widely used. When using this material, depending on the nature of the compounds included, sample pH adjustment prior to extraction is generally required. Less common cartridges used are Lichrolut ENV+, Oasis MCX, and StrataX. The fist is generally recommended for the extraction of polar organic compounds at low pH values, but it can also retain neutral compounds at pH 7 through hydrophobic interactions. Oasis MCX has been used to extract acidic, basic, and neutral compounds at low pH. Therefore, basic compounds are retained due to the cation-exchange properties and the acidic and neutral ones for the reversed-phase characteristics.

In the second, more selective, approach a class-specific extraction with molecular imprinted polymers (MIPs) can be used [43, 44]; however, their use in environmental analysis has been rarely reported and so far limited to pesticide analysis. For example, Turiel et al. [45] developed a method for group-selective extraction of chlorotriazines and methylyhiotyrizines and their TPs in soils, obtaining recoveries higher than 94% for chlorotriazines and 39% for prometryn. Similarly, in the study of Chapuis et al. [46], the ametryn MIP was shown to be highly class-selective for triazines and their degradation products and was applied to the clean-up of soil extracts.

Soils, sediments, and biosolids are complex matrices and extraction of organic contaminants and their TPs has been more challenging than in aqueous media since co-extracted material present in these samples can severely reduce the efficiency of extraction. Therefore, it is essential to develop effective methods for extraction and purification. Due to the thermolabile properties and polar nature of many TPs, traditional extraction methods such as Soxhlet are not appropriate and other techniques such as pressurized liquid extraction (PLE), microwave-assisted extraction (MAE), microwave-assisted Soxhlet extraction (MASE), and ultrasonic solvent extraction are more suitable. A summary of extraction methods applicable to the determination of pesticide TPs in solid matrices (soil) is presented in Table 4.

2.2.2
Detection and Quantitation

Although both the analytical techniques of GC and LC are applicable to the analysis of pesticides and their TPs, it is a common situation that parent pesticides can be analyzed by either GC or LC, whereas TPs can be analyzed only by LC because of their low volatility and their ionic properties. Pesticide transformation in the environment often results in very polar TPs, so that the polarity range to be covered by the chromatographic method must be extended, which makes GC analysis less suitable. In the case of polar compounds, LC coupled to MS and MS^2 is the method of choice, being triple quadrupole (QqQ), and the most widely used analytical tool in the analysis of TPs in environmental samples.

An example is analysis of chloroacetanilides and their TPs. Chloroacetanilides raised great concern due to evidence that their TPs might migrate into ground and surface water. They undergo transformation in the environment to form the corresponding ethansulfonic acids, which are typically encountered at concentrations much higher than the parent pesticides. Several analytical methods, based on LC-ESI-MS(/MS) in positive ion mode (for parent compounds) and negative ion mode (for ESA and OA transformation products), have been developed and validated for the determination of alachlor, acetochlor, metolachlor, dimethenamid, and their corresponding ESA and OA transformation products in water [47, 48] and soil [49].

Beside ESA and OA, a wide range of different neutral transformation products can form during biological, thermochemical, and photochemical transformations. However, limited information exists on their occurrence in the environment. In order to study the persistence of these neutral transformation products in coastal waters Hladik et al. [50] developed a GC-MS method for the analysis of 20 neutral chloroacetamide transformation products along with four parent compounds, three triazine herbicides, and two neutral triazine transformation products (Fig. 2). Using large volume injection, the obtained limits of detection were in the high picograms per liter or low nanograms per liter range. Table 5 lists analytical parameters for parent herbicides and neutral transformation products using GC-MS with selected ion monitoring.

Among PPCPs, the compounds that received the most attention are antibiotics, as they as they play a significant role in the increased emergence of antibiotic-resistant pathogenic bacteria [51]. Their analysis is typically performed using LC-MS, since their polarity, insufficient volatility, and thermal instability prevents their direct determination using GC. LC-MS has also become the technique of choice for the analysis of other pharmaceuticals. The technique has been used to study TPs of the β-blockers atenolol, metoprolol, and propranolol; the lipid-regulating agents atorvastatine and clofibric acid; psychiatric drugs carbamazepine and fluoxetine; and X-ray contrast media [52]. However, in most of the studies tandem MS or hybrid MS (i.e., Qq-TOF) instruments were used for identification and confirmation purposes, whereas only few papers reported on the development of quantitative methods for TPs in real environmental matrices.

An example is a method developed by Halling-Sorensen et al. [53] for the analysis of tetracyclines and their TPs in in soil interstitial water. Tetracyclines are often used in veterinary treatments and are frequently found in

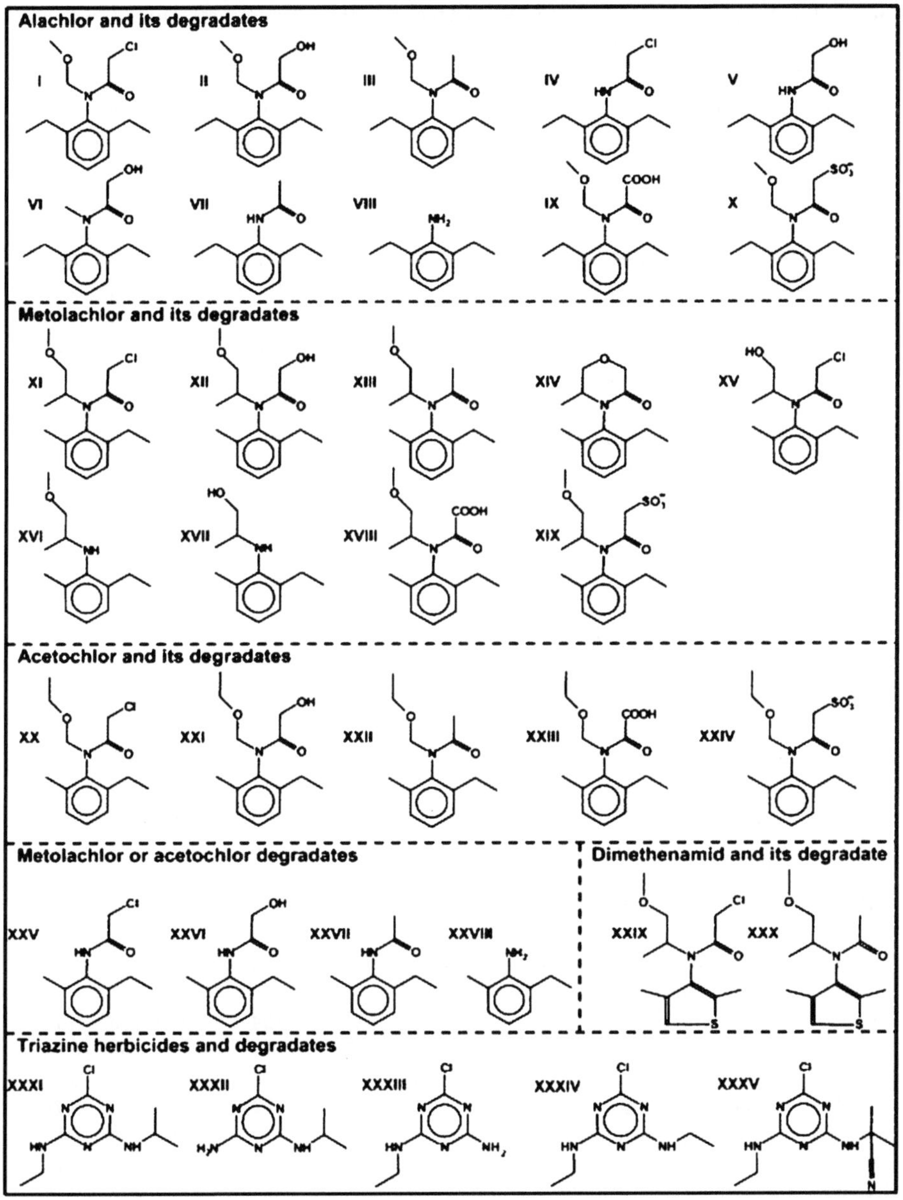

Fig. 2 Structures of parent herbicides and transformation products under investigation. Structures *I–X* represent alachlor and its transformation products; structures *XI–XIX* represent metolachlor and its transformation products; structures *XX–XXIV* represent acetochlor and its transformation products; structures *XXV–XXVIII* are those transformation products that can result from either metolachlor or acetochlor; structures *XXIX–XXX* represent dimethenamid and its transformation product; and structures *XXXI–XXXV* are triazine herbicides and their transformation products. Reprinted with permission from [50]

Table 5 Analytical parameters for GC-MS determination of parent herbicides and neutral degradates (adapted from [50])

Analyte[a]	Identity	MW	t_R[b] (min)	Quantitation ion	Monitoring ion(s)	MDL[c] (ng L^{-1})
I	Alachlor	269	21.55	160	237, 188	0.06
II	Hydroxyalachlor	251	20.49	160	219, 188	2.8
III	Deschloroalachlor	235	17.50	161	203, 178	0.24
IV	2-Chloro-2′-6′-diethylacetanilide	225	18.38	176	225, 148	0.13
V	2-Hydroxy-2′-6′-diethylacetanilide	207	21.1	176	207, 148	0.73
VI	2-Hydroxy-2′-6′-diethyl-*N*-methylacetanilide	221	18.86	190	221, 162	3.6
VII	2′-6′-Diethylacetanilide	191	16.64	148	191, 134	0.15
VIII	2,6-Diethylaniline	149	10.31	134	149, 119	–
XI	Metolachlor	283	22.72	162	238, 211	0.10
XII	Hydroxymetolachlor	265	21.73	162	220, 193	1.1
XIII	Deschlorometolachlor	249	18.62	162	204, 177	0.18
XIV	Metolachlor morpholinone	233	21.05–21.15	161	233, 188	0.15
XV	Metolachlor propanol	269	24.57	162	238, 146	0.18
XVI	Deschloroacetylmetolachlor	207	12.93	162	207, 133	0.10
XVII	Deschloroacetylmetolachlor propanol	193	15.36	162	193, 133	0.82
XX	Acetochlor	269	21.18	174	223, 162	0.15
XXI	Hydroxyacetochlor	251	20.11	174	205, 146	3.9
XXII	Deschloroacetochlor	235	17.04	164	206, 189	0.07
XXV	2-Chloro-2′-ethyl-6′-methyl-acetanilide	211	17.39	162	211, 134	0.22
XXVI	2-Hydroxy-2′-ethyl-6′-methyl-acetanilide	193	20.16	162	193, 134	0.80
XXVII	2′-Ethyl-6′-methyl-acetanilide	177	15.59	120	177, 134	0.19
XXVIII	2-Ethyl-6-methylaniline	135	8.88	120	135	–
XXIX	Dimethenamid	275	21.25	154	230, 203	0.10
XXX	Deschlorodimethenamid	241	17.11	154	196, 169	0.10
XXXI	Atrazine	215	19.41	200	215, 173	0.18
XXXII	Desethyl atrazine	187	18.14	172	187, 145	0.26
XXXIII	Desisopropyl atrazine	173	18.31	173	158, 145	0.15
XXXIV	Simazine	201	19.62	201	186, 173	0.20
XXXV	Cyanazine	240	24.48	225	240, 212	0.11

[a] Structures shown in Fig. 2
[b] Retention time
[c] Method detection limit for large-volume injection (100 μL)

Fig. 3 Chemical structures of OTC, EOTC, α-apo-OTC, β-apo-OTC, *N*-DM-OTC, E-*N*-DM-OTC, *N*-DDM-OTC, E-*N*-DDM-OTC, and ADOTC. From [53] ▶

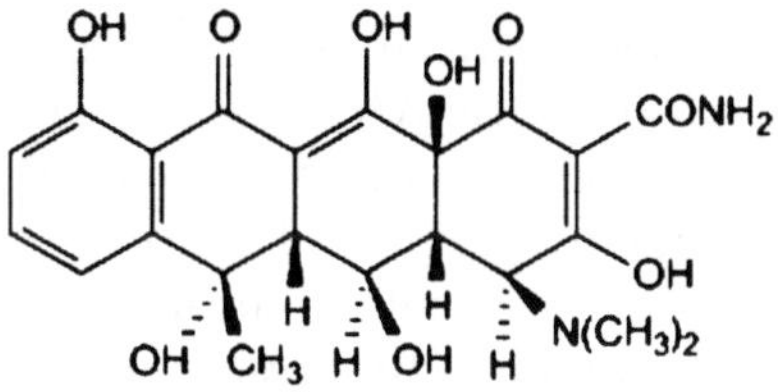

Oxytetracycline (OTC)

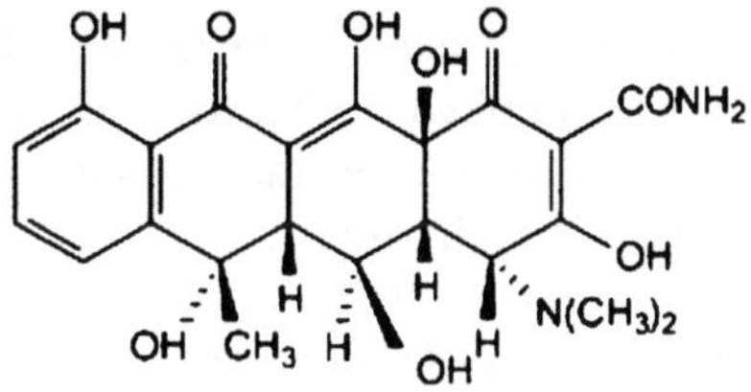

4-epi-oxytetracycline (EOTC)

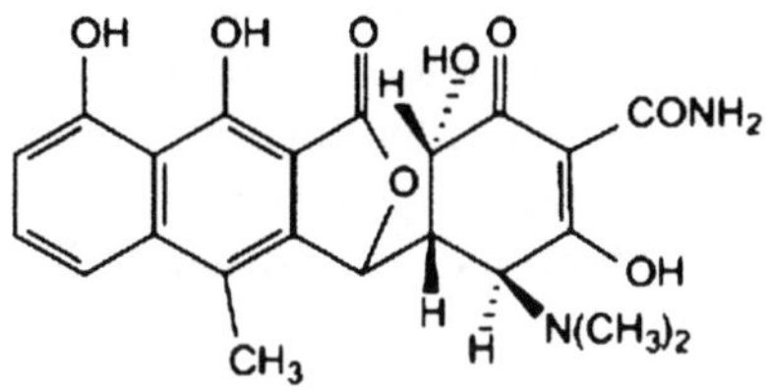

α-Apo-oxytetracycline (α-Apo-OTC)

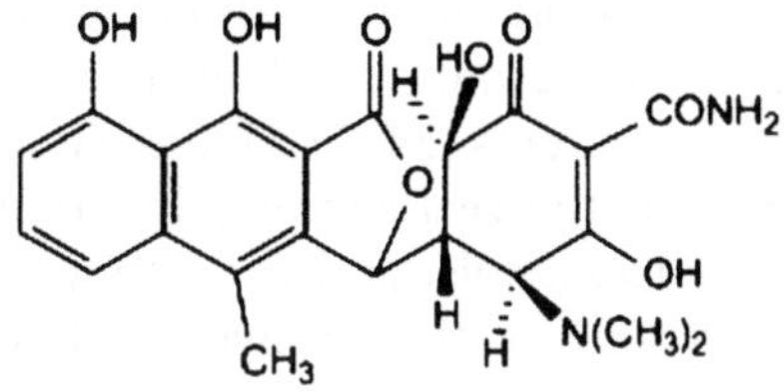

β-Apo-oxytetracycline (β-Apo-OTC)

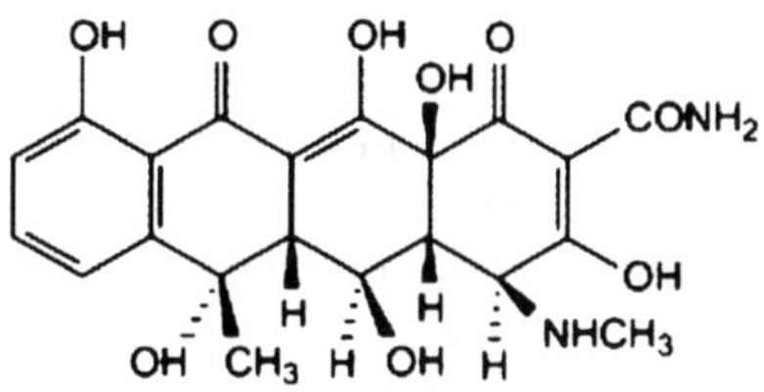

N-desmethyl-OTC (N-DM-OTC)

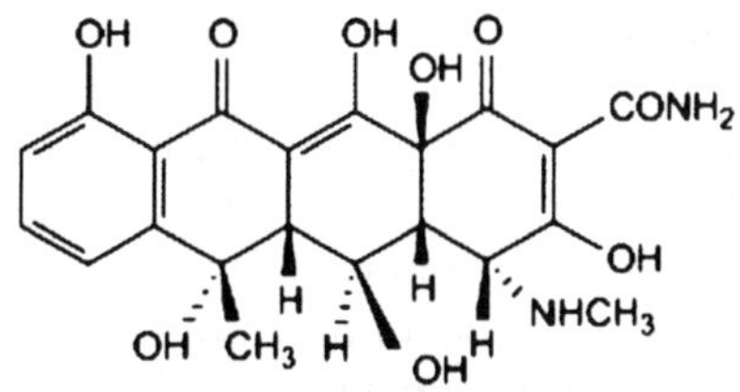

4-Epi-N-desmetyl-OTC (E-N-DM-OTC)

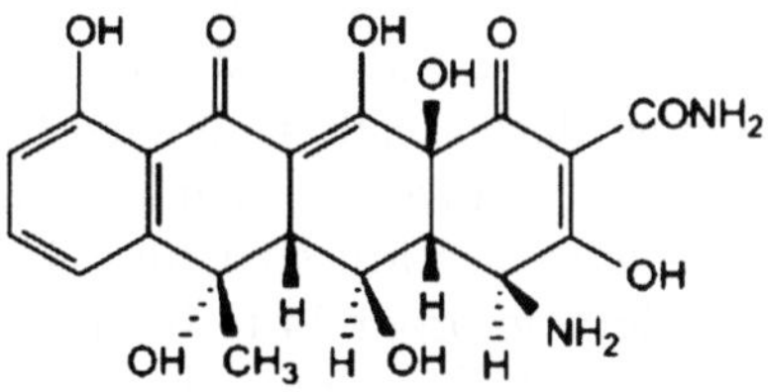

N-didesmethyl-OTC (N-DDM-OTC)

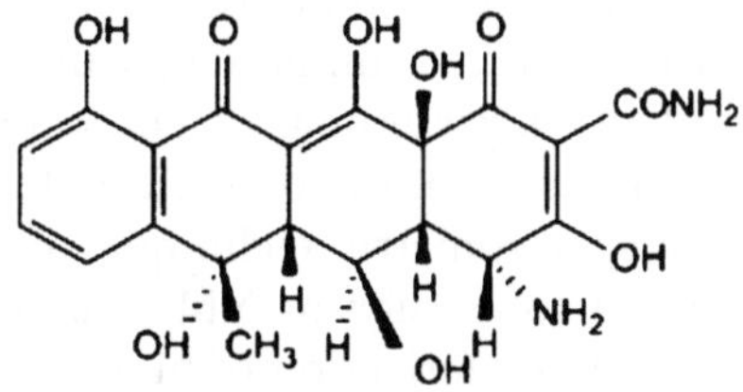

4-Epi-N-didesmethyl-OTC (E-N-DDM-OTC)

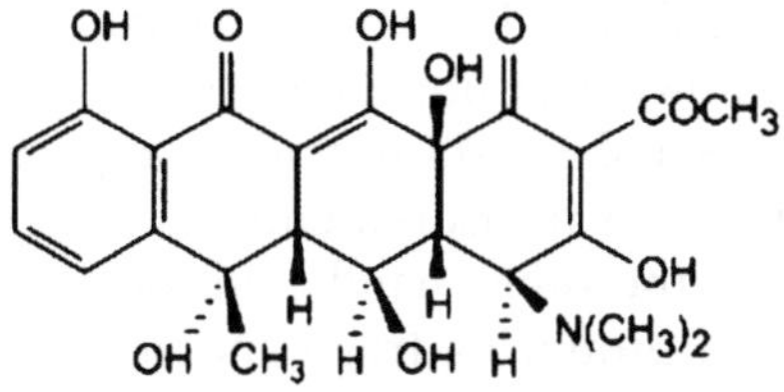

2-Acetyl-2-decarboxamido-oxytetracycline (ADOTC)

organic waste used as fertilizer. The method was based on LC-MS^2 (QqQ) and was applied to study the occurrence of oxytetracycline (OTC) and eight degradation products, namely 4-epi-OTC, α- and β-apo-OTC, 4-epi-*N*- and *N*-desmethyl-OTC, 4-epi-*N*- and *N*-didesmethyl-OTC, and 2-acetyl-2-decarboxamido-OTC in soil interstitial water (Fig. 3). Results indicated that OTC and 4-epi-OTC were the only compounds found to be present at significant concentrations in soil interstitial water, while all other degradation products were below 2% relative to OTC.

Several studies have shown that the photodegradation of triclosan, a commonly used antimicrobial agent, results in the generation of highly toxic photoproducts [54, 55]. Using GC-MS, several TPs were identified and quantified in the real samples including 2,8-dichlorodibenzo-*p*-dioxin (DCDD), dichlorophenols, and a compound tentatively identified as another DCDD congener or a dichlorohydroxydibenzofuran [56].

3 Conclusions

The presence of potentially harmful synthetic chemicals in the environment is of great concern as continuous exposure of wildlife and ecosystems to these chemicals may have deleterious effects. However, not only the compound in its unaltered form may pose a threat to environmental organisms, but transformation products of synthetic chemicals formed in natural or engineered settings need to be taken into consideration when aiming at constructing a more comprehensive picture of the contamination scenario. Since various processes are commonly involved in the conversion of xenobiotics in the environment, a complex mixture of parent compound and transformation products is generated. Its characterization is nowadays greatly facilitated by the employment of sophisticated analytical tools. The advent of hyphenated mass spectrometric techniques in combination with classical structure elucidation tools such as NMR has enabled considerable progress to be made in the characterization of biotransformation products of synthetic chemicals in complex environmental samples. In particular, the combination of high resolving ToF-MS and IT-MS instruments along with hybrid mass spectrometers such as QqToF-MS and QqLIT-MS have proved to be highly valuable techniques for studying unknown transformation products of synthetic chemicals. For example, ToF-MS and IT-MS provide complementary information; on the one hand accurate mass measurements allow elemental compositions to be proposed, while on the other hand multiple-stage experiments conducted in the IT-MS aid in proposing fragmentation pathways. Efficient and selective extraction methods allow concentration of trace levels of the intact parent compounds along with their transformation products present in complex environmental samples. Overall, the advances in instrumental analytical techniques, specifically in the development of mass spectrometric analyzers allowing the study of polar compounds, have provided tools of outstanding value, but undoubtedly there is still a tremendous gap to be bridged regarding the understanding of transformation processes and products of synthetic chemicals in the environment.

References

1. Petrović M, Hernando MD, Díaz-Cruz S, Barceló D (2005) J Chromatogr A 1067:1
2. Hopfgartner G, Varesio E, Tschäppät V, Grivet C, Bourgogne E, Leuthold LA (2004) J Mass Spectrom 39:845
3. Norwood DL, Qiu F (2004) Am Pharm Rev 7:92
4. Smyth WF (2003) Anal Chim Acta 492:1
5. Tozuka Z, Kaneko H, Shiraga T, Mitani Y, Beppu M, Terashita S, Kawamura A, Kagayama A (2003) J Mass Spectrom 38:793
6. Hernández F, Pozo OJ, Sancho JV, López FJ, Marín JM, Ibáñez M (2005) Trends Anal Chem 24:596
7. Reemtsma T (2003) J Chromatogr A 1000:477
8. Brown SC, Kruppa G, Dasseux JL (2005) Mass Spectrom Rev 24:223
9. Makarov A, Denisov E, Kholomeev A, Balschun W, Lange O, Strupat K, Horning S (2006) Anal Chem 78:2113
10. Prakash C, Shaffer CL, Nedderman A (2007) Mass Spectrom Rev 26:340
11. Fetzer JC, Biggs WR (1993) J Chromatogr 642:319
12. Biggs WR, Fetzer JC (2004) Anal Bioanal Chem 373:8
13. Günzler H, Gremlich HU (eds) (2002) IR spectroscopy: an introduction. Wiley-VCH, Weinheim
14. Cermola F, DellaGreca M, Iesce MR, Montanaro S, Previtera L, Temussi F (2006) Tetrahedron 62:7390
15. Class S (2005) Chem Eng News 83:15
16. Lovdahl MJ, Priebe SR (2000) J Pharm Biomed Anal 23:521
17. Chen Y, Rosazza JPN, Reese CP, Chang HY, Nowakowski MA, Kiplinger JP (1997) J Ind Microbiol Biotechnol 19:378
18. Pérez S, Eichhorn P, Barceló D (2007) Anal Chem 79:8293–8300
19. Raveton M, Aajoud A, Willison JC, Aouadi H, Tissut M, Ravanel P (2006) Environ Sci Technol 40:4151
20. Lam MW, Young CJ, Mabury SA (2005) Environ Sci Technol 39:513
21. Pérez S, Eichhorn P, Celiz MD, Aga DS (2006) Anal Chem 78:1866
22. Kalsch W (1999) Sci Total Environ 225:143
23. Quintana JB, Weiss S, Reemtsma T (2005) Water Res 39:2654
24. Sidelmann UG, Bjornsdottir I, Shockor JP, Hansen SH, Lindon JC, Nicholson JK (2001) J Pharm Biomed Anal 24:569
25. Wehrhan A (2006) Thesis. Rheinishen Friedrich-Wilhelms-Universität, Bonn
26. Grote M, Vockel A, Schwarze D, Mehlich A, Freitag M (2004) Fres Environ Bull 13:1214
27. Majewski MS, Foreman WT, Goolsby DA (2000) Sci Total Environ 248:201
28. Majewski MS, Foreman WT, Goolsby DA (2000) Sci Total Environ 248:213
29. Hernández F, Sancho JV, Pozo O, Lara A, Pitarch E (2001) J Chromatogr A 939:1
30. Sancho JV, Pozo O, Hernández F (2004) Analyst 129:38

31. Kolpin DW, Thurman EM, Goolsby DA (1996) Environ Sci Technol 30:335
32. Kolpin DW, Thurman EM, Linhart SM (2001) Environ Sci Technol 35:1217
33. Lerch RN, Thurman EM, Blanchard PE (1999) Environ Toxicol Chem 18:2161
34. Aelion CM, Mathur PP (2001) Environ Toxicol Chem 20:2411
35. Aga DS, Thurman EM (2001) Environ Sci Technol 35:2455
36. Aga DS, Thurman EM (2001) Environ Sci Technol 35:2455
37. Andreu V, Picó J (2004) Trends Anal Chem 23:772
38. Dabrowska D, Kot-Wasik A, Namiesnik J (2005) Crit Rev Anal Chem 35:155
39. Gros M, Petrovic M, Barcelo D (2006) Talanta 70:678
40. O'Connor S, Aga DS (2007) Trends Anal Chem 26:456
41. Hao C, Zhao X, Yang P (2007) Trends Anal Chem 26:569
42. Kot-Wasik A, Debska J, Namiesnik J (2007) Trends Anal Chem 26:557
43. Ensing K, Berggren C, Majors RE (2002) LCGC Eur 15:2
44. Suedee R, Srichana T, Chuchome T, Kongmark U (2004) J Chromatogr B 811:191
45. Turiel E, Maritn-Esteban A, Fernandez P, Pérez-Conde C, Camara C (2001) Anal Chem 73:5133
46. Chapuis F, Pichon V, Lanza F, Sellergren B, Hennion MC (2004) J Chromatogr B 804:93
47. Yokley RA, Mayer LC, Huang SB, Vargo JD (2002) Anal Chem 74:3754
48. Ferrer I, Thurman EM, Barcelo D (1997) Anal Chem 69:4547
49. Dignac T, Jeannot R, Mouvet C, Baran N (2002) J Chromatogr A 957:69
50. Hladik ML, Hsiao JI, Roberts AL (2005) Environ Sci Technol 39:6561
51. Chee-Sanford JC, Aminov RI, Krapac IJ, Garrigues-Jeanjean N, Mackie RI (2001) Appl Environ Microbiol 67:1494
52. Petrovic M, Barcelo D (2007) Trends Anal Chem 26:486
53. Halling-Sorensen B, Lykkeberg A, Ingerslev F, Nlackwell P, Tjornelund J (2003) Chemosphere 50:1331
54. Mezcua M, Gómez MJ, Ferrer I, Aguera A, Hernando MD, Fernández-Alba AR (2004) Anal Chim Acta 524:241
55. Aranami K, Readman JW (2007) Chemosphere 66:1052
56. Sanchez-Prado L, Llompart M, Lores M, Garciía-Jares C, Bayona JM, Cela R (2006) Chemosphere 65:1338
57. Lam MW, Mabury SA (2005) Aquat Sci 67:177
58. Chiron S, Minero C, Vione D (2006) Environ Sci Technol 40:5977
59. Eichhorn P, Aga DS (2004) Anal Chem 76:6002
60. Kwon JW, Armbrust KL (2005) Environ Toxicol Chem 24:1618
61. Doll TE, Frimmel FH (2003) Chemosphere 52:1757
62. Kalsch W (1999) Sci Total Environ 225:143
63. Haiß A, Kümmerer K (2006) Chemosphere 62:294
64. Galmier MJ, Bouchon B, Madelmont JC, Mercier F, Pilotaz F, Lartigue C (2005) J Pharm Biomed Anal 38:790
65. Agüera A, Pérez Estrada LA, Ferrer I, Thurman EM, Malato S, Fernández-Alba AR (2004) J Mass Spectrom 40:908
66. Edhlund BL, Arnold WA, McNeill K (2006) Environ Sci Technol 40:5422
67. Della Greca M, Iesce MR, Previtera L, Rubino M, Temussi F (2004) Environ Chem Lett 2:155
68. Winkler M, Lawrence JR, Neu TR (2001) Water Res 35:3197
69. Zwiener C, Seeger S, Glauner T, Frimmel FH (2002) Anal Bioanal Chem 372:569
70. Steger-Hartmann T, Länge R, Schweinfurth H, Tschampel M, Rehmann I (2002) Water Res 36:266
71. Bosca F, Miranda MA, Carganico G, Mauleon D (1994) Photochem Photobiol 60:96

72. Ternes TA, Kreckel P, Mueller J (1999) Sci Total Environ 225:91
73. Kiffmeyer T, Götze HJ, Jursch M, Lüders U (1998) Fres J Anal Chem 361:185
74. Moore DE, Chappuis PP (1998) Photochem Photobiol 47:173
75. Liu QT, Williams HE (2007) Environ Sci Technol 41:803
76. Boreen A, Arnold WA, McNeill K (2005) Environ Sci Technol 39:3630
77. Boreen A, Arnold WA, McNeill K (2004) Environ Sci Technol 38:3933
78. Lindström A, Buerge IJ, Bergqvist PA. Müller MD, Buser HR (2002) Environ Sci Technol 36:2322
79. Latch DE, Packer JL, Stender BL, VanOverbeke J, Arnold WA, McNeill K (2005) Environ Toxicol Chem 24:517
80. Ferrer I, Mezcua M, Gómez MJ, Thurman EM, Agüera A, Hernando MD, Fernández-Alba AR (2004) Rapid Commun Mass Spectrom 18:443
81. Eichhorn P, Ferguson PL, Pérez S, Aga DS (2005) Anal Chem 77:476

Hdb Env Chem Vol. 2, Part P (2009): 83–100
DOI 10.1007/698_2_011

Published online: 27 March 2008

Occurrence of Transformation Products in the Environment

Dana W. Kolpin[1] (✉) · William A. Battaglin[2] · Kathleen E. Conn[3] · Edward T. Furlong[4] · Susan T. Glassmeyer[5] · Steven J. Kalkhoff[1] · Michael T. Meyer[6] · Douglas J. Schnoebelen[1]

[1]U.S. Geological Survey, 400 South Clinton St., Iowa City, IA 52244, USA
dwkolpin@usgs.gov

[2]U.S. Geological Survey, Denver Federal Center, MS 415, Lakewood, CO 80225, USA

[3]Environmental Science and Engineering Division, Colorado School of Mines, 206 Coolbaugh Hall, Golden, CO 80401, USA

[4]U.S. Geological Survey, Denver Federal Center, MS 407, Lakewood, CO 80225, USA

[5]U.S. Environmental Protection Agency, National Exposure Research Laboratory, 26 West Martin Luther King Drive, MS 564, Cincinnati, OH 45268, USA

[6]U.S. Geological Survey, 4821 Quail Crest Place, Lawrence, KS 66049, USA

Abstract Historically, most environmental occurrence research has focused on the parent compounds of organic contaminants. Research, however, has documented that the environmental transport of chemicals, such as pesticides and emerging contaminants, are substantially underestimated if transformation products are not considered. Although most examples described herein were drawn from research conducted by the U.S. Geological Survey, such results are generally reflective of those found in other parts of the world. Results from a study of 51 streams in the Midwestern United States found that transformation products were seven of the ten most frequently detected pesticide compounds in late spring runoff (after application of pre-emergent herbicides), and nine of the ten most frequently detected compounds in fall season runoff (during and after harvest). In fact, 70% of the total herbicide concentration in water from the Mississippi River Basin was from transformation products. Results from a study of 86 municipal wells in Iowa found the frequency of detection increased from 17%, when pesticide parent compounds were considered, to 53%, when both parents and transformation products were considered. Transformation products were 12 of the 15 most frequently detected compounds for this groundwater study. Although studies on transformation products of synthetic organic compounds other than pesticides are not as common, wastewater treatment plant discharges have repeatedly been shown to contribute such transformation products to streams. In addition, select detergent transformation products have been commonly found in solid waste in the 1000's mg/kg. These findings and many others document that transformation products must be considered to fully assess the potential environmental occurrence of chemical contaminants and their transport and fate in various compartments of the hydrologic system.

Keywords Ground water · Surface water · Transformation products

1 Introduction

Historically, most environmental occurrence research has focused on the parent compounds of organic contaminants [1]. Research has repeatedly shown, however, that transformation products of many organic compounds, such as pesticides [2–6], pharmaceuticals [7, 8], personal care products [9, 10], surfactants [11–13], and hormones [14, 15], are present in the environment, including less studied environmental compartments, such as estuaries [16], air [17], precipitation [18–20], and plant and animal tissue [21–23]. Such transformation products are often present in higher environmental concentrations than their parent compounds [24–26]. Thus, environmental occurrence of synthetic organic compounds can be substantially underestimated when transformation products are ignored. Although substantial progress has been made towards environmental monitoring of transformation products, analytical methods to measure and detect many of these compounds have still not been developed.

The purpose of this chapter is not to provide a complete literature review on the topic of the environmental occurrence of transformation products, but rather to provide specific examples of research documenting the importance of including transformation products in investigations of environmental contaminants. Although most examples described herein were drawn from research conducted by the U.S. Geological Survey, such results are generally reflective of those found in other parts of the world.

2 Pesticides

Numerous studies worldwide have documented the presence of pesticide transformation products in the environment [27–32]. The below investigations are provided as examples documenting the importance of including transformation products for pesticide studies of both streams and groundwater.

2.1 Streams

During 2002, 154 samples were collected from 51 streams in nine states across the Midwestern United States during three periods of runoff [33, 34]. The three sampling periods were all within one growing season and translated into runoff generally following the application of pre-emergent pesticides (May–June), runoff generally following the application of post-emergent pesticides (June–July), and runoff during or after harvest (September–November). All samples were analyzed for 21 pesticide parent compounds and 27 pesticide transformation products using gas chromatography/mass spectrometry (GC/MS) and liquid chromatography/mass spectrometry (LC/MS) techniques [35–38]. Pesticide transformation products were seven of the ten most frequently detected compounds in runoff samples collected early in the growing season (following the application of pre-emergent pesticides) and nine of the ten most frequently detected compounds in runoff samples collected at the end of the growing season (Fig. 1A and B). In general, parent compounds were present in higher concentrations than transformation products during late-spring runoff, but transformation product concentrations tended to be higher than their parent compounds during harvest-season runoff [34]. For example, the median ratio of AMPA (the primary glyphosate degradation product) to glyphosate was 0.85 in pre-emergent samples, 1.29 in post-emergent samples, and 1.87 in harvest season samples [33].

In a study that included comprehensive spatial and temporal sampling, 70% of the total herbicide concentration in water from the Mississippi River Basin was from transformation products [39]. Other studies in the United States have documented similar results. For example, in 1993, the annual load of alachlor ethanesulfonic acid was almost 500 metric t; whereas, the annual load of alachlor was less than 50 metric t [40].

Schnoebelen et al. (2003) [41] found wide seasonal fluctuations in total concentrations for parent pesticide compounds with season highs corresponding to time of application and runoff events. Conversely, total concentrations of their transformation products remained relatively constant temporally and were generally greater in concentration than the parent compounds (Fig. 2). Consequently, transport of transformation products from the watershed to streams occurs more constantly over time, whereas transport of the parent product is more directly influenced by episodic application and seasonal hydrology. This pattern was present in both the headwaters and the mouth of the river with much of the yearly pesticide load transported to the Mississippi River being triazine and chloroacetanilide transformation products [28]. A study of 70 Midwestern streams sampled during base-flow conditions found total concentrations of transformation products (4.4 μg/L) to be almost 17 times greater than the total concentrations of the parent

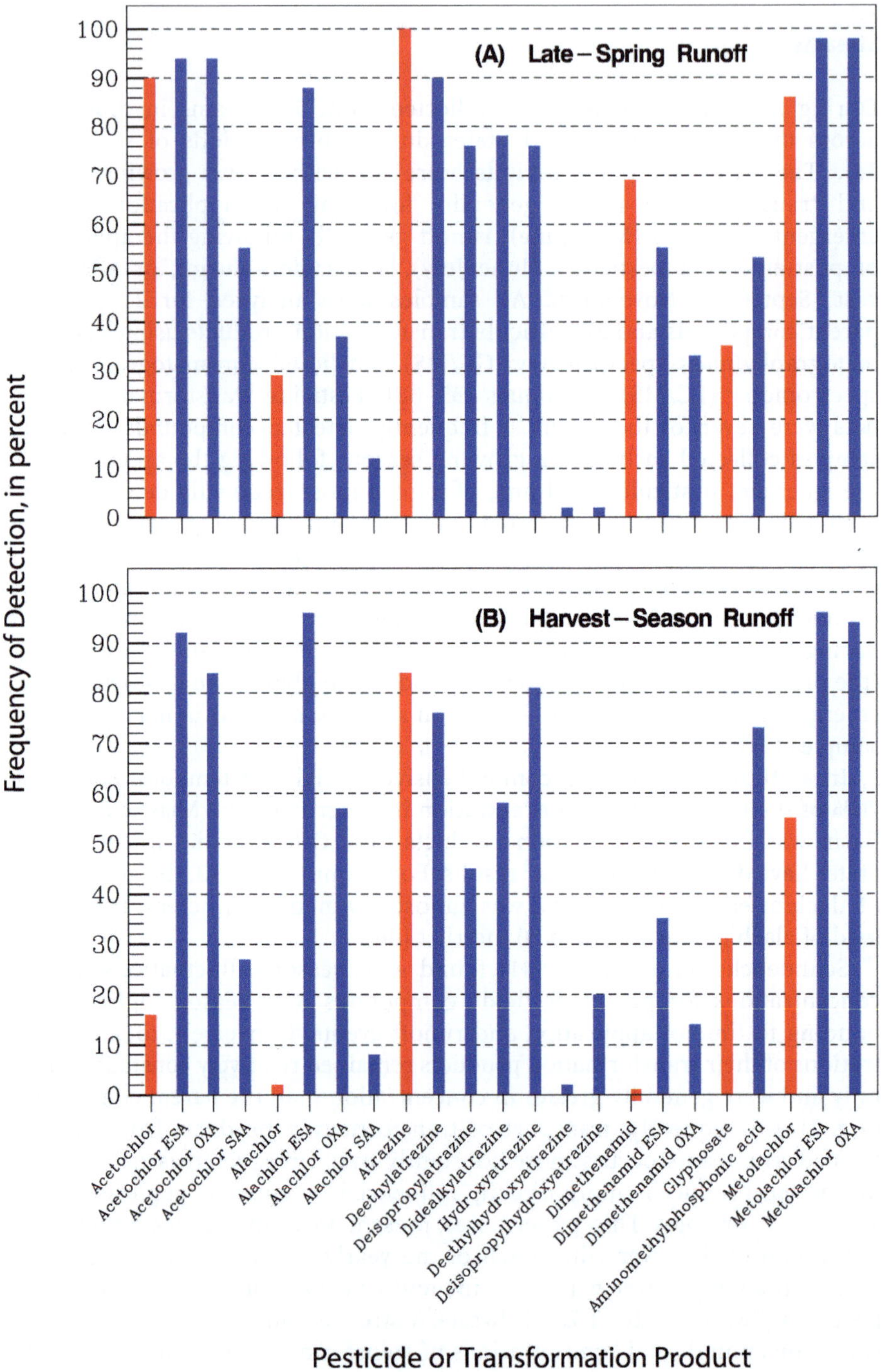
(A) Late – Spring Runoff
(B) Harvest – Season Runoff
Frequency of Detection, in percent
0
10
20
30
40
50
60
70
80
90
100
Acetochlor
Acetochlor ESA
Acetochlor OXA
Acetochlor SAA
Alachlor
Alachlor ESA
Alachlor OXA
Alachlor SAA
Atrazine
Deethylatrazine
Deisopropylatrazine
Didealkylatrazine
Hydroxyatrazine
Deethylhydroxyatrazine
Deisopropylhydroxyatrazine
Dimethenamid
Dimethenamid ESA
Dimethenamid OXA
Glyphosate
Aminomethylphosphonic acid
Metolachlor
Metolachlor ESA
Metolachlor OXA
Pesticide or Transformation Product

◄ **Fig. 1** Frequency of detection of herbicides (*red bars*) and their transformation products (*blue bars*) in samples collected from 51 US Midwestern streams during late spring runoff (**A**) and harvest season runoff (**B**) [34]. ESA – ethane sulfonic acid; OXA – oxanilic acid; SAA – sulfinyl acetic acid

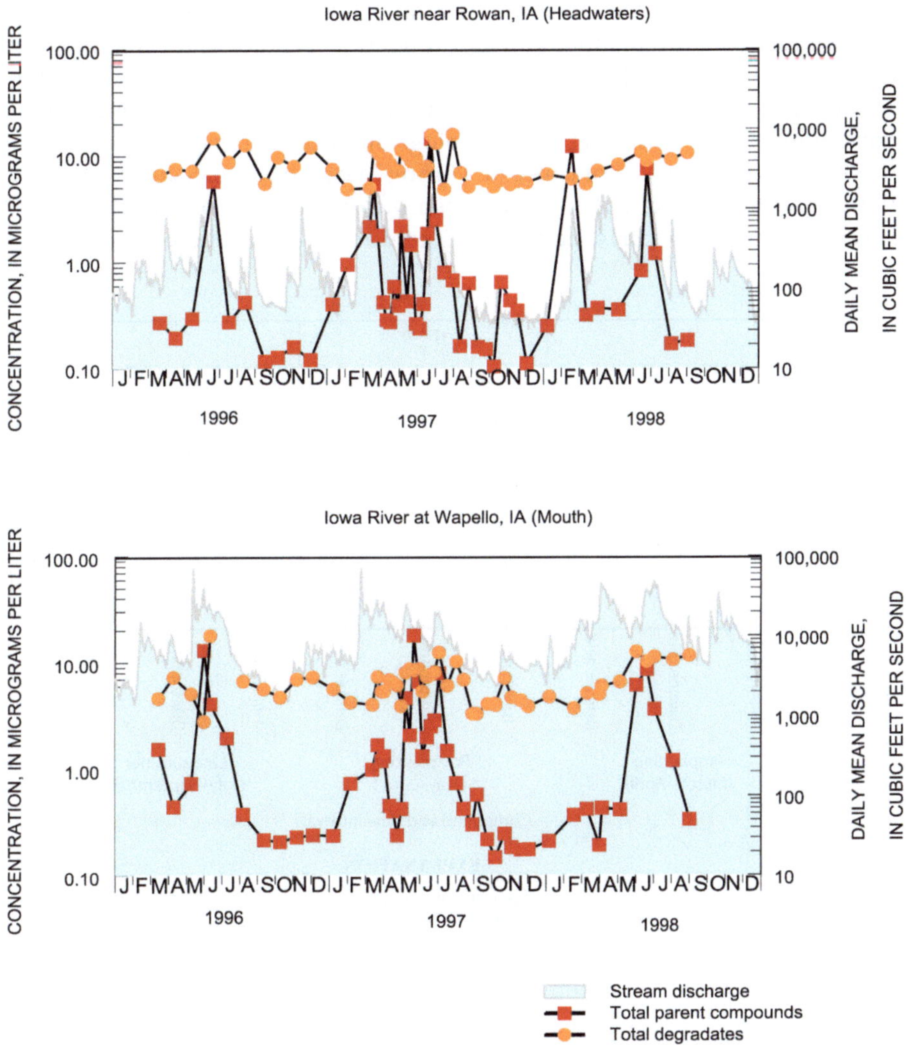

Fig. 2 Seasonal variability of pesticides and pesticide transformation products near the headwaters and the mouth of the Iowa River, 1996–98 [41]. Total parent compound concentration was the sum of detected concentrations for 21 agricultural herbicides, three nonagricultural herbicides, and nine insecticides. Total transformation product concentration was the sum of detected concentrations for seven chloroacetanilide transformation products and four triazine transformation products

compounds (0.26 μg/L) [42]. This is not surprising because base flow is sustained by groundwater in these streams, and therefore, the residence time and opportunity for transformation of pesticides is enhanced as they are transported from field, to aquifer, and, finally, to streams [43].

Isoxaflutole is a newer, low-application rate preemergent herbicide that, while extensively used on corn in the Unites States, has little information available on its environmental occurrence. Isoxaflutole is unique in that, by the design of the manufacturer, it quickly degrades to the biologically active diketonitrile [44] followed by transformation of diketonitrile to the herbicidally inactive benzoic acid. Diketonitrile is more water soluble than isoxaflutole. Thus, transport of the biologically active transformation product to the environment is of concern. During March through September 2004, 75 water samples were collected from ten major rivers in Iowa to determine the environmental occurrence of isoxaflutole and two isoxaflutole transform-

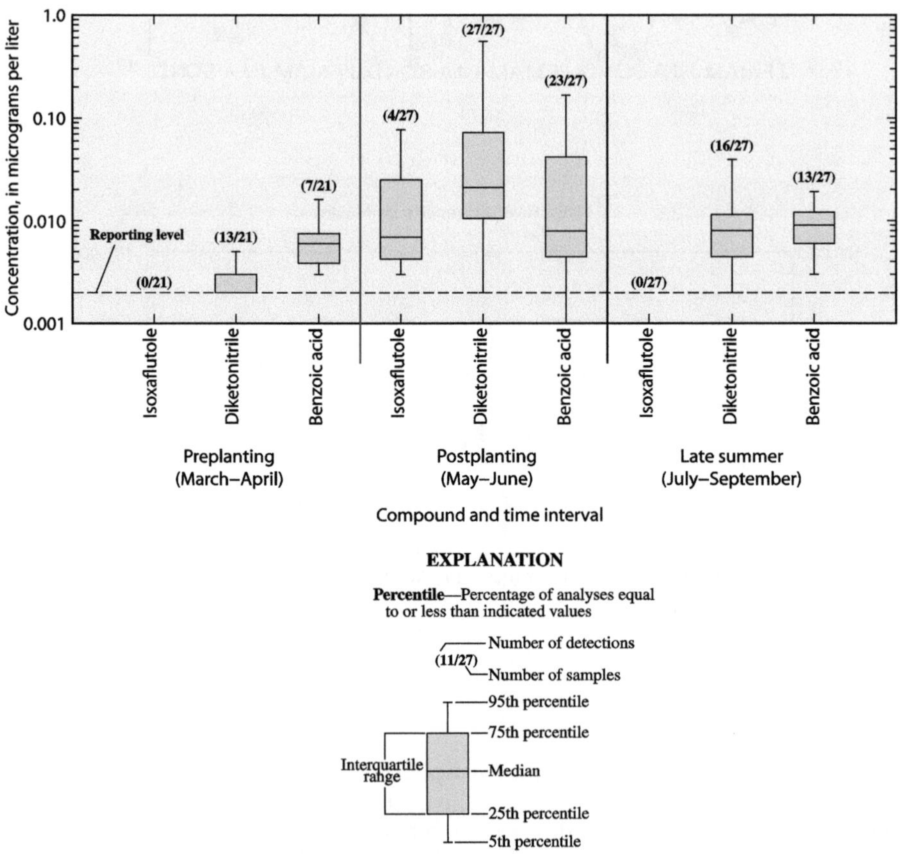

Fig. 3 Concentrations of isoxaflutole, diketonitrile, and benzoic acid in water samples collected from ten rivers in Iowa, 2004 [45]

ation products [45, 46] using vacuum manifold solid-phase extraction and LC/MS [47]. Only 5% of the stream samples contained measurable concentrations of isoxaflutole (maximum of 0.077 μg/L), while 75% contained the biologically active transformation product diketonitrile (maximum of 0.552 μg/L) and 57% contained benzoic acid (maximum of 0.166 μg/L) (Fig. 3). The largest number of detections and concentrations of all three isoxaflutole compounds occurred during and shortly after the planting season. These results document that transformation products can be important even for low-application rate pesticides.

2.2 Groundwater

During 2001, a network of 86 municipal wells was sampled in Iowa representing groundwater from all major aquifer types (alluvial, bedrock/karst region, glacial drift, bedrock/nonkarst region) [48]. All samples were analyzed for 21 pesticide parent compounds and 24 pesticide degradates using GC/MS and LC/MS techniques [35–38]. The frequency of detection increased from 17%, when only parent compounds were considered, to 53%, when both parents and transformation products were considered, with transformation products being 12 of the 15 most frequently detected compounds (Fig. 4).

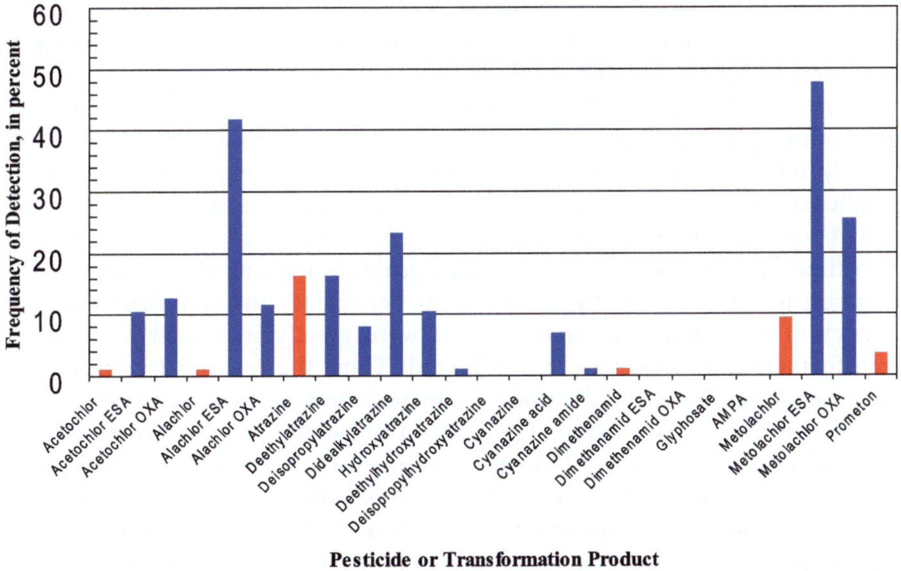

Fig. 4 Frequency of detection for selected pesticides (*red bars*) and their transformation products (*blue bars*) from 86 municipal wells sampled in Iowa during 2001 [48]. ESA – ethane sulfonic acid; OXA – oxanilic acid; AMPA – aminomethyl phosphonic acid

Table 1 Summary of detection frequencies for select pesticide parent compounds and parent compounds plus transformation products for 86 wells sampled in 2001 [48]

Compound	Frequency of Detection (%) Parent	Parent plus transformation products
Acetochlor	1.2	16.3
Alachlor	1.2	44.2
Atrazine	15.3	29.1
Cyanazine	0	18.6
Dimethenamid	1.2	1.2
Flufenacet	0	0
Glyphosate	0	0
Metolachlor	9.3	48.8
Any pesticide	17.4	53.5

These results are similar to a study of Illinois groundwater, where parent compounds were detected in 11% and transformation products were detected in 56% of the wells sampled [49]. Other research has shown that concentrations of transformation products are typically higher than corresponding parent compounds at study sites with thick unsaturated zones, but not at sites with thin unsaturated zones, suggesting that degradation occurs primarily in the unsaturated zone for many pesticides [50]. For most parent compounds, the transport of pesticide compounds is substantially underestimated when transformation products are not considered, even for a relatively persistent compound, such as atrazine (Table 1). Interestingly, however, glyphosate and its transformation product aminomethyl phosphonic acid (AMPA) were not found during this groundwater study [48], even though glyphosate use has rapidly increased since 1998, with the introduction of genetically altered glyphosate-resistant crops [51]. The lack of detections of glyphosate and AMPA for this study is likely due to the combination of insufficient transport time for these compounds to reach groundwater and their affinities for sorption to sediment particles. For most pesticides, however, both the frequency of detection and concentration were greatly increased when their transformation products were considered. Thus, the transport of pesticide compounds was substantially underestimated when pesticide transformation products were not considered. The greatest underestimates were determined for alachlor and cyanazine [48], pesticides which had little use during the time of sample collection, but had substantial use during the 10–15 years prior to sampling. Previous research has shown that pesticides and pesticide transformation products can continue to be present in ground water years after the active ingredient had last been used [52].

3 Emerging Contaminants

Many chemicals which have not historically been considered environmental contaminants are being found with increasing frequency in the environment on a global scale [11, 53–60]. These "emerging contaminants" (ECs) are commonly derived from municipal, agricultural, and industrial wastewater sources and pathways and include a diverse set of contaminants including pharmaceuticals, personal care products, and synthetic and biogenic hormones. These newly recognized contaminants represent a shift in traditional thinking, as many are produced industrially and yet are dispersed to the environment from domestic, commercial, and industrial uses. As analytical methods become available, similar to results found for pesticides, transformation products of these ECs are also being documented in the environment worldwide [25, 60–69]. In fact, it was the prevalence of clofibric acid, an active metabolite of the blood lipid regulators clofibrate, etofibrate, and etofylliclofibrate [70], in European water resources [71–73] that was a major factor in spurring the plethora of pharmaceutical-related studies currently being conducted. The below investigations are provided as examples documenting the importance of including EC transformation products for various types of EC studies of both waste sources and streams.

3.1 Waste Sources

Recent research has begun to characterize the presence of ECs in a variety of waste sources (e.g., wastewater treatment plants, onsite septic systems, etc.) to better understand their potential pathways into the environment. Although data on transformation products for emerging contaminants is more limited when compared to other compounds, such as pesticides, a growing number of methods to detect such transformation products are being developed and select compounds have been found in a variety of waste sources, including wastewater treatment plants [7, 67, 74–78], septic systems [13, 79], landfills [80], and animal manure [81, 82]. For example, in a study of waste from wastewater treatment plants [75] and septic systems [13] (Tables 2–3), 1,7-dimethylxanthine (caffeine metabolite), 4-nonylphenol diethoxylate and 4-nonylphenol monoethoxylate (transformation products of surfactants commonly used in domestic detergents), and cotinine (nicotine metabolite) occur in wastewater treatment plant and onsite septic system effluents at concentrations typically less than 100 μg/L. In general, concentrations of transformation products were much higher in solid waste when compared to liquid waste for both wastewater treatment plants (Table 2) and onsite septic systems (Table 3). Concentrations of select surfactant transformation products were commonly in the 1000's μg/kg with maximum concentrations reaching much higher

Table 2 Summary of analytical results of select emerging contaminants from a study of 11 wastewater treatment plant effluents [75] and 15 wastewater treatment plant biosolids [83]. RL – reporting level; Freq Detec – frequency of detection; Max – maximum concentration; ND – not detected

Compound	Use	Treated municipal effluent			Treated municipal biosolid		
		RL (μg/L)	Freq Detec (%)	Max (μg/L)	RL (μg/kg)	Freq Detect (%)	Max (μg/kg)
Caffeine	Stimulant	0.016	72.7	7.99	1.3	93.3	1200
1,7-Dimethylxanthine	Caffeine metabolite	0.14	36.4	8.55	2.0	60.0	6200
4-Cumylphenol	Detergent metabolite	1.0	0.0	ND	33.7	13.3	5030
4-Nonylphenol diethoxylate	Detergent metabolite	5.0	90.9	38	852	100	4600
4-Nonylphenol monoethoxylate	Detergent metabolite	5.0	90.9	18	336	100	25 300
4-*tert*-Octylphenol	Detergent metabolite	1.0	36.4	1.1	22.9	100	2400
Cotinine	Nicotine metabolite	0.023	90.9	1.03	1.3	60.0	45
Dehydronifedipine	Antianginal metabolite	0.015	72.7	0.021	1.7	33.3	26
Erythromycin-H_2O	Erythromycin metabolite	0.02	72.7	0.48	1.7	33.3	41
para-Nonylphenol-total	Detergent metabolite	5.0	27.3	22	499	100	483 000

Table 3 Summary of analytical results of select emerging contaminants in five onsite septic systems in Colorado [13]. RL, reporting level; Freq Detec, frequency of detection; Max, maximum concentration; NM, not measured; ND, not detected

Compound	Use	Liquid waste RL (μg/L)	Liquid waste Freq Detec (%)	Liquid waste Max (μg/L)	Solid waste RL (μg/kg)	Solid waste Freq Detect (%)	Solid waste Max (μg/kg)
Caffeine	Stimulant	0.016	100	88	NM	NM	NM
1,7-Dimethylxanthine	Caffeine metabolite	0.14	100	56	NM	NM	NM
4-Cumylphenol	Detergent metabolite	1	0	ND	25	0	ND
4-Nonylphenol diethoxylate	Detergent metabolite	5	60.0	35	250	80.0	22 000
4-Nonylphenol monoethoxylate	Detergent metabolite	NM	NM	NM	250	100	44 000
4-*tert*-Octylphenol	Detergent metabolite	1	40.0	2.1	25	80.0	23 000
Cotinine	Nicotine metabolite	0.014	100	3.9	NM	NM	NM
Dehydronifedipine	Antianginal metabolite	0.015	0	ND	NM	NM	NM
Erythromycin-H_2O	Erythromycin metabolite	0.005	80.0	0.2	NM	NM	NM
para-Nonylphenol-total	Detergent metabolite	5	80.0	38	250	100	1 800 000

levels. Caffeine is an example of where both the parent compound and a transformation product (1,7-dimethylxanthine) were measured in waste samples. Caffeine was almost ubiquitous in waste sources with 1,7-dimethylxanthine also being commonly detected, sometimes in higher concentrations than the parent compound (Tables 2–3). Previous research has shown that a transformation product of sulfamethoxazole (N^4-acetylsulfamethoxazole) accounted for up to 86% of the total sulfamethoxazole load in untreated wastewater [84]. Other parent EC compounds have been found to be more persistant, however, with transformation products of oxytetracycline accounting for only 5% of the total oxytetracycline load in treated effluent [77].

3.2 Streams

The first national-scale study of emerging contaminants in the United States found that such compounds were commonly present in stream systems, with 80% of the streams sampled having at least one compound detected [11]. Over 60% of the chemical concentrations exceeding 1 μg/L were from three detergent metabolites (4-nonylphenol, 4-nonylphenol monoethoxylate, and 4-nonylphenol diethoxylate). In addition, transformation products also were frequently detected (e.g., cotinine, 1,7-dimethylxanthine, erythromycin H_2O). This study indicated the enormous potential for parent compounds to degrade during environmental transport and for the transformation products to contribute much of the relative loading of these contaminants. The results confirm the need to include both parent compounds and transformation products in occurrence studies.

Other studies [60, 62, 75, 77] also clearly documented the contributions of wastewater treatment plant effluents to concentrations of transformation products in streams, with chemical concentrations and frequency of detections significantly greater ($P<0.05$, Mann–Whitney U test) downstream from the wastewater treatment plant when compared to samples collected upstream from the wastewater treatment plant [75]. In a study of select wastewater treatment plants and receiving streams in the United Kingdom, acetyl-sulfamethoxazole was detected almost four times more frequently than its parent compound (33% versus 9%) in the final effluents sampled, with only acetyl-sulfamethoxazole being detected in the corresponding receiving streams [62]. Additional research has shown that concentrations of emerging contaminants and their transformation products in urban-impacted streams decrease with increasing streamflow [85]. Thus, the urban contribution becomes progressively muted as streamflow increases.

Based on the current understanding of organic environmental chemistry, as well as supporting studies, it is clear that only the most recalcitrant chemicals will persist untransformed in the environment under common ambient conditions. Therefore, it is not surprising that transformation products of various ECs, such as caffeine, detergents, and other synthetic organic compounds (e.g., pesticides) will substantially contribute to the environmental loads of such contaminants. It is also likely that a broad range of pharmaceutical transformation products will continue to be identified in the environment as analytical methods become more available and subsequent environmental occurrence studies are undertaken (e.g. [63, 86]).

Acknowledgements Author's would like to thank Mike Focazio (USGS) and Tom Behymer (USEPA) for their helpful comments during the technical review process. This paper has been reviewed in accordance with the peer and administrative review policies of the U.S. Geological Survey and the U.S. Environmental Protection Agency. The use of trade, product, or firm names is for descriptive purposes only and does not imply endorsement by the U.S. Government.

References

1. Boxall ABA, Sinclair CJ, Fenner K, Kolpin DW, Maund SJ (2004) When synthetic chemicals degrade in the environment. Environ Sci Technol 38:369A–375A
2. Kolpin DW, Thurman EM, Gooslby DA (1996) Occurrence of selected pesticides and their metabolites in near-surface aquifers of the Midwestern United States. Environ Sci Technol 30:335–340
3. Gooddy DC, Chilton PJ, Harrison I (2002) A field study to assess the degradation and transport of diuron and its metabolites in a calcareous soil. Sci Total Environ 297:67–83
4. Battaglin WA, Thurman EM, Kalkhoff SJ, Porter SD (2003) Herbicides and transformation products in surface waters of the Midwestern United States. J Am Water Res Assoc 39:743–756
5. Baran N, Mouvet C, Dagnac T, Jeannot R (2004) Infiltration of acetochlor and two of its metabolites in two contrasting soils. J Environ Qual 33:241–249
6. Hladik ML, Hsiao JJ, Roberts AL (2005) Are neutral chloroacetamide herbicide degradates of potential environmental concern? Analysis and occurrence in the upper Chesapeake Bay. Environ Sci Technol 39:6561–6574
7. Miao X-S, Yang J-J, Metcalfe CD (2005) Carbamazepine and its metabolites in wastewater and in biosolids in a municipal wastewater treatment plant. Environ Sci Technol 39:7469–7475
8. Søeberg T, Ingerslev F, Halling-Sørensen B (2004) Chemical stability of chlortetracycline and chlortetracycline degradation products and epimers in soil interstitial water. Chemos 57:1515–1524
9. Bester K (2005) Fate of triclosan and triclosan-methyl in sewage treatment plants and surface waters. Arch Environ Contam Toxicol 49:9–17
10. Aranami K, Readman JW (2007) Photolytic degradation of triclosan in freshwater and seawater. Chemos 66:1052–1056
11. Kolpin DW, Furlong ET, Meyer MT, Thurman EM, Zaugg SD, Barber LB, Buxton HT (2002) Pharmaceuticals, hormones, and other organic wastewater contaminants in US streams, 1999–2000: A national reconnaissance. Environ Sci Technol 36:1202–1211
12. Cespedes R, Lacorte S, Raldua D, Ginebreda A, Barcelo D, Pina B (2005) Distribution of endocrine disruptors in the Llobregat River Basin (Catalonia, NE Spain). Chemos 61:1710–1719
13. Conn KE, Barber LB, Brown GK, Siegrist RL (2006) Occurrence and fate of organic contaminants during onsite wastewater treatment. Environ Sci Technol 40:7358–7366

14. Chen C-Y, Wen T-Y, Wang G-S, Cheng H-W, Lin Y-H, Lien G-W (2007) Determining estrogenic steroids in Taipei waters and removal in drinking water treatment using high-flow solid-phase extraction and liquid chromatography/tandem mass spectrometry. Sci Total Environ 378:352–365
15. Fernandez MP, Ikonomou MG, Buchanan I (2007) An assessment of estrogenic organic contaminants in Canadian wastewaters. Sci Total Environ 373:250–269
16. McConnell LL, Rice CP, Hapeman CJ, Drakeford L, Harmon-Fetcho JA, Bialek K, Fulton MH, Leight AK, Allen G (2007) Agricultural pesticides and selected degradation products in five tidal regions and the main stem of Chesapeake Bay USA. Environ Toxicol Chem 26:2567–2578
17. Foreman WT, Majewski MS, Goolsby DA, Wiebe FW, Coupe RH (2000) Pesticides in the atmosphere of the Mississippi River Valley, part II - air. Sci Total Environ 248:213–216
18. Bucheli TD, Muller SR, Heberle S, Schwarzenbach RP (1998) Occurrence and behavior of pesticides in rainwater, roof runoff, and artificial stormwater infiltration. Environ Sci Technol 32:3457–3464
19. Dorfler U, Scheunert I (1997) S-triazine herbicides in rainwater with special reference to the situation in Germany. Chemos 35:77–85
20. Goolsby DA, Thurman EM, Pomes ML, Meyer MT, Battaglin WA (1997) Herbicides and their metabolites in rainfall–Origin, transport, and deposition patterns across the midwestern and northeastern United States, 1990–1991. Environ Sci Technol 31:1325–1333
21. Brooks BW, Chambliss CK, Stanley JK, Ramirez A, Banks KE, Johnson RD, Lewis RJ (2005) Determination of select antidepressants in fish from an effluent-dominated stream. Environ Toxicol Chem 24:464–469
22. Correa-Reyes G, Viana MT, Marquez-Rocha FJ, Licea AF, Ponce E, Vazquez-Duhalt R (2007) Nonylphenol algal bioaccumulation and its effect through the trophic chain. Chemos 68:662–670
23. Rice CP, Schmitz-Afonso I, Loyo-Rosales JE, Link E, Thoma R, Fay L, Altfater D, Camp MJ (2003) Alkylphenol and alkylphenol-ethoxylates in carp, water, and sediment from the Cuyahoga River, Ohio. Environ Sci Technol 37:3747–3754
24. Kolpin DW, Thurman EM, Linhart SM (2000) Finding minimal herbicides concentrations in ground water? Try looking for their degradates. Sci Total Environ 248:115–122
25. Gasser L, Fenner K, Scheringer M (2007) Indicators for the exposure assessment of transformation products of organic micropollutants. Environ Sci Technol 41:2445–2451
26. Hladik ML, Bouwer EJ, Roberts AL (2008) Neutral chloroacetamide herbicide degradates and related compounds in Midwestern United States drinking water sources. Sci Total Environ 390:155–165
27. Galassi S, Provini A, Mangiapan S, Benfenati E (1996) Alachlor and its metabolites in surface water. Chemos 32:229–237
28. Kalkhoff SJ, Barnes KK, Becher KD, Savoca ME, Schnoebelen DJ, Sadorf EM, Porter SD, Sullivan DJ (2000) Water quality in the Eastern Iowa Basins, Iowa and Minnesota, 1996–98. US Geological Survey Circular 1210
29. Papastergiou A, Papadopoulou-Mourkidou P (2001) Occurrence and spatial and temporal distribution of pesticide residues in groundwater of major corn-growing areas of Greece (1996–1997). Environ Sci Technol 35:63–69
30. Sarmah AK, Muller K, Ahmad R (2004) Fate and behavior of pesticides in the agroecosystem - a review with a New Zealand perspective. Austr J Soil Res 42:125–154
31. Gilliom RJ, Barbash JE, Crawford CG, Hamilton PA, Martin JD, Nakagaki N, Nowell LH, Scott JC, Stackelberg PE, Thelin GP, Wolock DM (2006) Pesticides in the

nation's streams and ground water, 1992–2001: The quality of our nation's water. U.S. Geological Survey Circular 1291. (http://ca.water.usgs.gov/pnsp/pubs/circ1291/)
32. Hildebrandt A, Lacorte S, Barcelo D (2007) Assessment of priority pesticides, degradation products, and pesticide adjuvants in groundwaters and top soils from agricultural areas of the Ebro river basin. Anal Bioanal Chem 387:1459–1468
33. Battaglin WA, Kolpin DW, Scribner E, Kuivila KK, Sandstrom MA (2005) Glyphosate, other herbicides, and transformation products in Midwestern streams, 2002. J Am Water Res Assoc 41:323–332
34. Scribner EA, Battaglin WA, Dietze JE, Thurman EM (2003) Reconnaissance data for glyphosate, other selected herbicides, their degradation products, and antibiotics in 51 streams in nine midwestern States, 2002. U.S. Geological Survey Open-File Report 03–217
35. Zimmerman LR, Thurman EM (1998) Method of analysis by the U.S. Geological Survey Organic Geochemistry Group – Determination of triazine and chloroacetanilide herbicides in water by solid-phase extraction and capillary column gas chromatography/mass spectrometry with selected ion monitoring. U.S. Geological Survey Open-File Report 98–634
36. Zimmerman LR, Schneider RJ, Thurman EM (2002) Analysis and detection of the herbicides dimethenamid and flufenacet and their sulfonic and oxanilic acid degradates in natural water. J Agri Food Chem 50:1045–1052
37. Lee EA, Kish JL, Zimmerman LR, Thurman EM (2001) Methods of analysis by the U.S. Geological Survey Organic Geochemistry Research Group – Update and additions to the determination of chloroacetanilide herbicide degradation compounds in water using high performance liquid chromatography/mass spectrometry. U.S. Geological Survey Open-File Report 01–10
38. Lee EA, Strahan AP, Thurman EM (2001) Methods of analysis by the U.S. Geological Survey Organic Geochemistry Research Group – Determination of glyphosate, aminomethylphosphonic acid, and glufosinate in water using online solid-phase extraction and high-performance liquid chromatography/mass spectrometry. U.S. Geological Survey Open-File Report 01–454
39. Rebich RA, Coupe RH, Thurman EM (2004) Herbicide concentrations in the Mississippi River Basin – the importance of chloroacetanilide herbicide degradates. Sci Total Environ 321:189–199
40. Clark GM, Goolsby DA, Battaglin WA (1999) Seasonal and annual load of herbicides from the Mississippi River Basin to the Gulf of Mexico. Environ Sci Technol 33:981–986
41. Schnoebelen DJ, Kalkhoff SJ, Becher KD, Thurman EM (2003) Water-quality assessment of the Eastern Iowa Basins: Selected pesticides and pesticide degradates in streams, 1996–98. U.S. Geological Survey Water-Resources Investigations Report 03–4075
42. Kalkhoff SJ, Lee KE, Porter SD, Terrio PJ, Thurman EM (2003) Herbicides and herbicide degradation products in upper Midwest agricultural streams during August base-flow conditions. J Environ Qual 32:1025–1035
43. Squillace PJ (1996) Observed and simulated movement of bank-storage water. Ground Water 34:121–134
44. Lin CH, Lerch RN, Thurman EM, Garrett HE, George MF (2002) Determination of isoxaflutole (Balance) and its metabolites in water using solid phase extraction followed by high-performance liquid chromatography with ultraviolet or mass spectrometry. J Ag Food Chem 50:5816–5824
45. Scribner EA, Meyer MT, Kalkhoff SJ (2006) Occurrence of isoxaflutole, acetamide, and triazine herbicides and their degradation products in 10 Iowa rivers draining to the

Mississippi and Missouri Rivers, 2004. U.S. Geological Survey Scientific Investigations Report 2006–5169
46. Meyer MT, Scribner EA, Kalkhoff SJ (2007) Comparison of fate and transport of isoxaflutole to atrazine and metolachlor in 10 Iowa Rivers. Environ Sci Technol 41:6933–6939
47. Meyer MT, Lee EA, Scribner EA (2007) Methods of analysis by the U.S. Geological Survey Organic Geochemistry Research Group–Determination of dissolved isoxaflutole and its sequential degradation products, diketonitrile and benzoic acid in water using solid phase extraction and liquid chromatography/mass spectrometry. U.S. Geological Survey Techniques and Methods, book 5, chap. A9, p 13
48. Kolpin DW, Schnoebelen DJ, Thurman EM (2004) Degradates provide insight to spatial and temporal trends of herbicides in ground water. Ground Water 42:601–608
49. Mills PC, Kolpin DW, Scribner EA, Thurman EM (2004) Herbicides and degradates in shallow aquifers of Illinois – Spatial and temporal trends. J Am Water Res Assoc 41:537–547
50. Tesoriero AJ, Saad DA, Burow KR, Frick EA, Puckett LJ, Barbash JE (2007) Linking ground-water age and chemistry along flow paths: Implications for trends and transformations of nitrate and pesticides. J Contam Hydrol 94:139–155
51. Padgette SR, Kolacz KH, Delannay X (1995) Development, identification, and characterization of a glyphosate-tolerant soybean line. Crop Sci 35:1451–1461
52. Morvan X, Mouvet C, Baran N, Gutierrez A (2006) Pesticides in the groundwater of a spring draining a sandy aquifer: Temporal variability of concentrations and fluxes. J Contam Hydrol 87:176–190
53. Bruchet A, Hochereau C, Picard C, Decottignies V, Rodrigues JM, Janex-Habibi ML (2005) Analysis of drugs and personal care products in French source and drinking waters: the analytical challenge and examples of application. Water Sci Technol 52:53–61
54. Gulkowska A, He Y, So MK, Yeung LWY, Leung HW, Giesy JP, Lam PKS, Martin M, Richardson BJ (2007) The occurrence of selected antibiotics in Hong Kong coastal waters. Marine Poll Bull 54:287–1306
55. Halling-Sørensen B, Nielson SN, Lanzky PF, Ingerslev F, Holten Lutzhoft J, Jorgensen SE (1998) Occurrence, fate and effects of pharmaceutical substances in the environment – A review. Chemos 35:357–393
56. Heberer T (2002) Occurrence, fate, and removal of pharmaceutical residues in the aquatic environment: A review of the recent research data. Toxicol Letters 131:5–17
57. Kim SD, Cho J, Kim IS, Vanderford BJ, Snyder SA (2007) Occurrence and removal of pharmaceutical and endocrine disruptors in South Korean surface, drinking, and waste waters. Water Research 41:1013–1021
58. Moldovan Z (2006) Occurrences of pharmaceutical and personal care products as micropollutants in rivers from Romania. Chemos 64:1808–1817
59. Ternes TA (1998) Occurrence of drugs in German sewage treatment plants and rivers. Water Res 32:3245–3260
60. Wiegel S, Aulinger A, Brockmeyer R, Harms H, Loffler J, Reincke H, Schmidt R, Stachel B, von Tumpling W, Wanke A (2004) Pharmaceuticals in the river Elbe and its tributaries. Chemos 57:107–126
61. Lindstrom A, Buerge IJ, Poiger T, Bergqvist PA, Muller MD, Buser HR (2002) Occurrence and environmental behavior of the bactericide triclosan and its methyl derivative in surface waters and in wastewater. Environ Sci Technol 36:2322–2329
62. Ashton D, Hilton M, Thomas KV (2004) Investigating the environmental transport of human pharmaceuticals to streams in the United Kingdom. Sci Total Environ 333:167–184

63. Chiron S, Minero C, Vione D (2006) Photodegradation processes of the antiepileptic drug carbamazepine, relevant to estuarine waters. Environ Sci Technol 40:5977–5983
64. Matamoros V, Bayona JM (2006) Elimination of pharmaceuticals and personal care products in subsurface flow constructed wetlands. Environ Sci Technol 40:5811–5816
65. Benotti MJ, Brownawell BJ (2007) Distributions of pharmaceuticals in an urban estuary during both dry- and wet-weather conditions. Environ Sci Technol 41:5795–5802
66. Buth JM, Arnold WA, McNeill K (2007) Unexpected products and reaction mechanisms of the aqueous chlorination of cimetidine. Environ Sci Technol 41:6228–6233
67. Horii Y, Reiner JL, Loganathan BG, Kumar KS, Sajwan K, Kannan K (2007) Occurrence and fate of polycylic musks in wastewater treatment plants in Kentucky and Georgia USA. Chemos 68:2011–2020
68. Managaki S, Murata A, Takada H, Tuyen BC, Chiem NH (2007) Distribution of macrolides, sulfonamides, and trimethoprim in tropical waters: Ubiquitous occurrence of veterinary antibiotics in the Mekong Delta. Environ Sci Technol 41:8004–8010
69. Martin C, Moeder M, Daniel X, Krauss G, Schlosser D (2007) Biotransformation of the polycyclic musks HHCB and AHTN and metabolite formation by fungi occurring in freshwater environments. Environ Sci Technol 41:5395–5402
70. Stumpf M, Ternes TA, Wilken RD, Rodrigues SV, Baumann W (1999) Polar drug residues in sewage and natural waters in the state of Rio de Janeiro, Brazil. Sci Total Environ 225:135–141
71. Franke S, Hildebrandt S, Schwarzbauer J, Link M, Francke W (1995) Organic compounds as contaminants of the Elbe river and its tributaries Part II: GC/MS screening for contaminants of the Elbe water. Fresenius J Anal Chem 353:39–49
72. Heberer T, Butz S, Stan HJ (1995) Analysis of phenoxycarboxylic acids and other acidic compounds in tap, ground, surface and sewage water at the low ng/l level. Int J Environ Anal Chem 58:43–53
73. Buser HR, Muller MD, Theobald N (1998) Occurrence of the pharmaceutical drug clofibric acid and the herbicide mecoprop in various Swiss lakes and the North Sea. Environ Sci Technol 32:188–192
74. Clara M, Strenn B, Gans O, Martinez E, Kreuzinger N, Kroiss H (2005) Removal of selected pharmaceuticals, fragrances and endocrine disrupting compounds in a membrane bioreactor and conventional wastewater treatment plants. Water Res 39:4797–4807
75. Glassmeyer ST, Furlong ET, Kolpin DW, Cahill JD, Zaugg SD, Werner SL, Meyer MT, Kryak DD (2005) Transport of chemical and microbial compounds from known wastewater discharges: Potential for use as indicators of human fecal contamination. Environ Sci Technol 39:5157–5169
76. Gobel A, McArdell CS, Joss A, Siegrist H, Giger W (2007) Fate of sulfonamides, macrolides, and trimethoprim in different wastewater treatment technologies. Sci Total Environ 372:361–371
77. Li D, Yang M, Hu J, Ren L, Zhang Y, Li K (2008) Determination and fate of oxytetracycline and related compounds in oxytetracycline production wastewater and the receiving river. Environ Toxicol Chem 27:80–86
78. Ying G-G, Kookana RS, Kumar A (2008) Fate of estrogens and xenoestrogens in four sewage treatment plants with different technologies. Environ Toxicol Chem 27:87–94
79. Godfrey E, Woessner WW, Benotti MJ (2007) Pharmaceuticals in on-site sewage effluent and ground water, western Montana. Ground Water 45:263–271
80. Barnes KK, Christenson SC, Kolpin DW, Focazio MJ, Furlong ET, Zaugg SD, Meyer MT, Barber LB (2004) Pharmaceuticals and other organic waste water contam-

inants within a leachate plume downgradient of a municipal landfill. Ground Water Monit Remed 24:119–126
81. Hanselman TA, Graetz DA, Wilkie AC (2003) Manure-borne estrogens as potential environmental contaminants: A review. Environ Sci Technol 37:5471–5478
82. Hutchins SR, White MR, Hudson FM, Fine DD (2007) Analysis of lagoon samples from different concentrated animal feeding operations for estrogens and estrogen conjugates. Environ Sci Technol 41:738–744
83. Kinney CA, Furlong ET, Zaugg SD, Burkhardt MR, Werner SL, Cahill JD, Jorgensen GR (2006) Survey of organic wastewater contaminants in biosolids destined for land application. Environ Sci Technol 40:7207–7215
84. Gobel A, Thomsen A, Mcardell CS, Joss A, Giger W (2005) Occurrence and sorption behavior of sulfonamides, macrolides, and trimethoprim in activated sludge treatment. Environ Sci Technol 39:3981–3989
85. Kolpin DW, Skopec M, Meyer MT, Furlong ET, Zaugg SD (2004) Urban contribution of pharmaceuticals and other organic wastewater contaminants to streams during differing flow conditions. Sci Total Environ 328:119–130
86. Lishman L, Smyth SA, Sarafin K, Kleywegt S, Toito J, Peart T, Lee B, Servos M, Beland M, Seto P (2006) Occurrence and reductions of pharmaceuticals and personal care products and estrogens by municipal wastewater treatment plants in Ontario, Canada. Sci Total Environ 367:544–558

Part II
Exposure of Transformation Products

Hdb Env Chem Vol. 2, Part P (2009): 103–120
DOI 10.1007/698_2_018

Published online: 13 May 2009

Fate of Transformation Products of Synthetic Chemicals

Dingfei Hu (✉) · Keri Henderson · Joel Coats

Pesticide Toxicology Laboratory, Department of Entomology, Iowa State University, Ames, IA 50011, USA
dingfei-hu@uiowa.edu

Abstract With increasing utilization of various synthetic chemicals, the adverse impact of these chemicals is of concern due to their occurrence in the environment. The subsequent transformation products pose considerable risk on the ecosystem and human health. However, for most currently used synthetic chemicals, the fate of their transformation products are yet to be elucidated, which forces most current risk assessment to focus on parent chemicals. In this article, general processes and principles of environmental fate of chemical transformation products are illustrated; various influential factors such as physiochemical properties of transformation products and conditions of the environmental media are described; a mass balance approach used to investigate the environmental fate is discussed.

Keywords Atrazine and other synthetic chemicals · Fate · Transformation products

1 Introduction

Pesticides, pharmaceuticals, cosmetics, food additives, and industrial chemicals are among the tens of thousands of synthetic chemicals being utilized currently in our society. Every chemical we use has the potential to persist in the surroundings, and any transformation products formed by biotic or abiotic processes are then found in the environment. Synthetic chemicals and their transformation products have reached almost every corner of the world, with real and potential impact on humans, wildlife, and plant life on the earth. Even in the polar regions, traces of synthetic chemicals or their transformation products, including polychlorinated naphthalenes (PCNs), polychlorinated dibenzo-*p*-dioxins (PCDDs), polychlorinated dibenzofurans (PCDFs), polychlorinated biphenyls (PCBs), and pesticides (HCB, *p*,*p*′-DDE), have been found in polar organisms such as the polar bear from the Alaskan Arctic and krill, sharp-spined notothen, crocodile icefish, Antarctic silverfish, Adelie penguin, South polar skua, and Weddell seal from the Ross Sea, Antarctica [1]. It is speculated that these chemicals were transported there by ocean currents, atmospheric movement, or other animals.

To assess the potential hazards of chemicals, most risk assessments target primarily persistent parent compounds. However, the environmental exposure results from not only parent chemicals but also their transformation products [2]. Compared to their parent compounds, chemical transformation products can have longer or shorter persistence, greater, equal or less toxicity, and possible interaction with other substances ([3–6]; see Sinclair and Boxall, in this volume). Therefore, it is critical that parent chemicals and their transformation products are all evaluated during ecological risk assessments. Although the environmental significance of chemical transformation products has long been realized since the discovery of the negative environmental impact of two DDT transformation products (DDE and DDD) decades ago, the fate of transformation products of most synthetic chemicals are yet to be investigated due to the lack of analytical methods and available analytical standards [6, 7]. Herein we review literature and studies to shed some light on the environmental fate of transformation products of synthetic chemicals.

2 Processes

Once introduced into the environment, a chemical is influenced by many processes, which determine its persistence, transport, and ultimate destination. Through abiotic or biotic degradation, transformation products form in the environment. The chemical and its transformation products can move vertically through the soil profile to groundwater and away from the source site with mobile groundwater. They also have the potential to reach surface water when they travel laterally either as surface runoff or through subsoil tile drains, entering streams, major rivers, reservoirs, and ultimately estuaries and oceans. Various microbes, plants, animals and even humans might

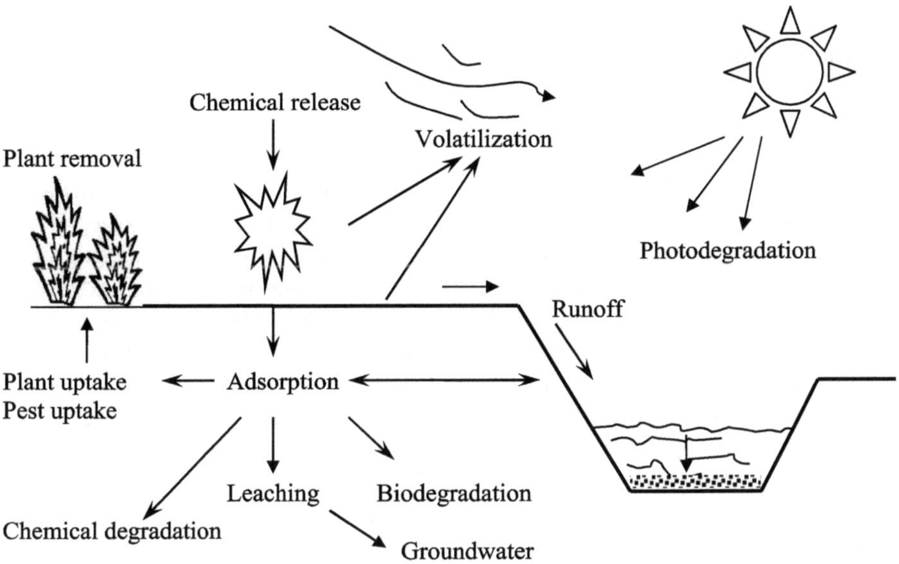

Fig. 1 Environmental fate processes of a released chemical and its transformation products

take up and biodegrade the chemicals to some extent. Some compounds move into the atmosphere through volatilization, adsorption onto dust, or just as solid particles. The possible fate processes for a synthetic chemical and its transformation products in the environment are illustrated in Fig. 1. These processes apply to both parent compounds and their subsequent transformation products formed in the environment, although those transformation products might not necessarily behave in the same way as their parent forms.

These processes, chemical degradation, biodegradation, and photodegradation, affect persistence, and other processes, such as binding, uptake into plants and animals, volatilization, runoff, leaching, all influence the movement of a chemical and its transformation products in the environment. Chemical transformation products can possibly undergo any of the described processes at any point during the whole course of the environmental fate of a chemical. For example, trichloroacetic acid (TCA) is a major atmospheric transformation product formed by photo-oxidation of chlorocarbons, which are used in large amounts industrially as solvents and herbicides. TCA has higher polarity than chlorocarbons, and can deposit with precipitation as a contaminant in water and soil, and further degrades into carbon dioxide in biota such as microbes and plants [8]. In spite of formation of different transformation products during these processes, most synthetic chemicals are ultimately degraded into carbon dioxide, mineral salts, water, and humic substances in the environment.

2.1
Transformation

Chemicals are usually degraded via chemical, biological, or physical means through a multi-step process in the environment (see Wackett et al., in this volume). Figure 2 shows an example of degradation pathways of the broadly used herbicide atrazine [9]. Hydrolysis of atrazine to hydroxyatrazine (HYA) is the primary abiotic degradation pathway. Dealkylation of atrazine is the main biodegradation pathway to form deethylatrazine (DEA), and some species of microorganisms form deisopropylatrazine (DIA) by deisopropylation. All intermediate transformation products are further degraded, and then eventually are mineralized into inorganic compounds such as CO_2, H_2O or NH_3 in the environment. Atrazine is one pesticide which has been extensively investigated from the perspective of the fate of both the parent compound and its transformation products. Therefore, atrazine and its transformation products are taken as examples herein to elucidate the behavior of transformation products in the environment.

In general, transformation products have less persistence in the environment than the parent compounds. However, some transformation products from a range of chemical classes including certain carbamates, triazines, organophosphates, and sulfonylureas have often shown more persistence than the parent forms (Fig. 3) [3]. Meanwhile, it was pointed out that these observations might be skewed because of biased reports of more persistent transformation products and limitations of experimental design such as decreased microbial activity in the test system [3]. Differences between persistence of transformation products and their parent compounds exist for sure; therefore, case-by-case study of each chemical seems necessary to investigate the potential impact of the transformation products.

2.2
Biodegradation

In reality, transformation products and their parent compounds often coexist in the environment. Sometimes the properties of transformation products might induce biodegradation due to induction of enzymes or inhibit biodegradation due to their toxicity to soil microbes responsible for degradation [10]. Enhanced biodegradation is the result of adaptation of a microbial community which has been previously exposed to a chemical, and the adapted microorganisms subsequently use this chemical as a primary energy or nutrient source [11]. Whether chemicals are toxic or serving as nutrients, they must be available to microbes to exert these actions. Therefore, transformation products with low toxicity, high nutritive value, and bioavailability may allow microorganisms to thrive, in particular ones that can utilize the molecule as a nutritive substrate. On the other hand, transformation products

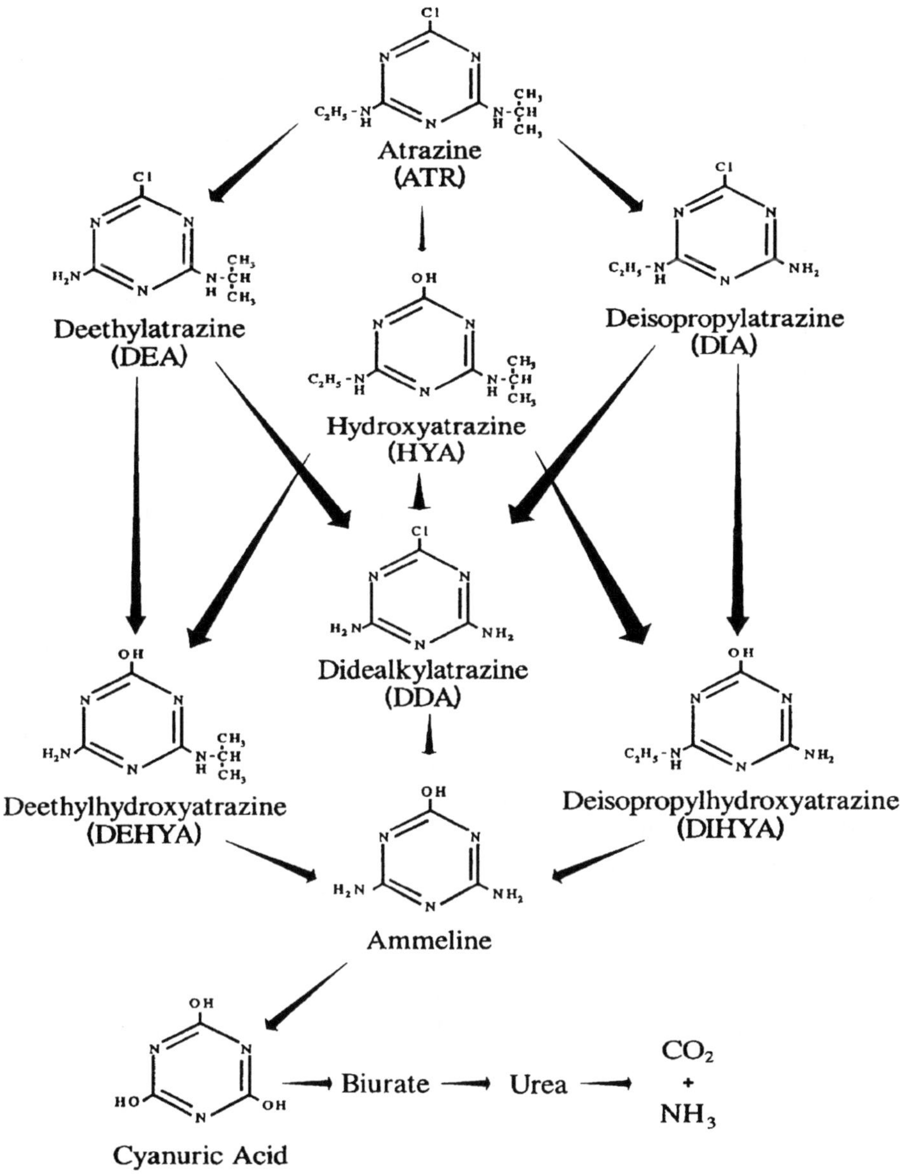

Fig. 2 Degradation of atrazine in the environment (Reproduced from [9], ©1993 SETAC)

might inhibit further degradation. PCBs are degraded aerobically via a four-step process referred to as the *bph* pathway, but enzymes responsible for each step can produce metabolites that inhibit subsequent enzymes to function properly, resulting in incomplete degradation [12].

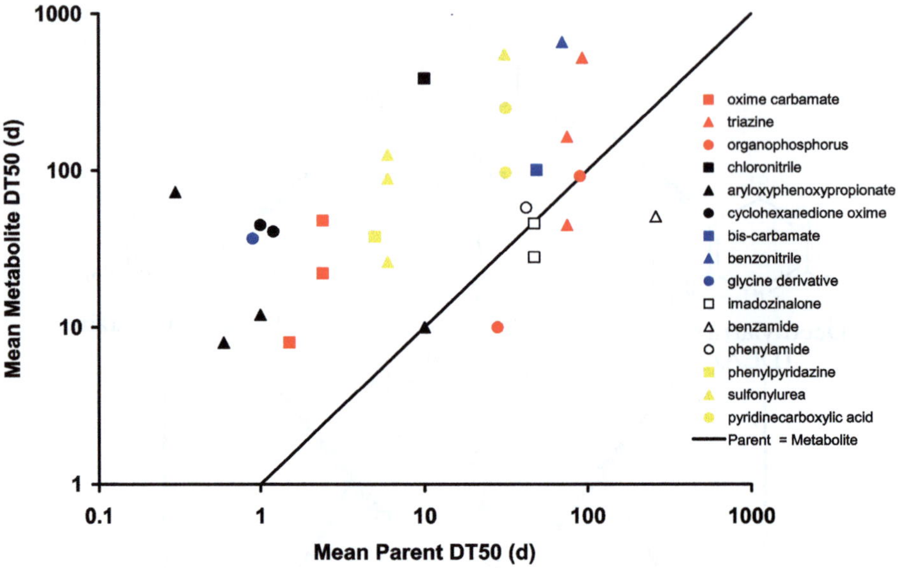

Fig. 3 The half-lives of transformation products of a range of pesticides in soil are longer than those of their parents. (Reproduced with permission from [3])

The presence of chemical transformation products has shown rapid degradation of their parent compounds [9, 13–16]. Repeated application of some pesticides can also accelerate the degradation of the formed transformation products in the environment [17], and some examples of such transformation products and respective parent compounds are listed in Table 1. Carbendazim, 2-aminobenzimidazole, methyl isothiocyanate, and desmethyldiphenamid were degraded faster after repeated exposure of the respective parent compounds – carbendazim, benomyl, diphenamid, and metham-sodium [18–21]. DEA degradation was enhanced in soils with long-term exposure to atrazine (atrazine-history soil) compared to soils

Table 1 Transformation products susceptible to enhanced degradation after soil has been treated with their parent compounds

Transformation products	Parent compounds	Refs.
Desmethyldiphenamid	Metham-sodium	[18]
2-Aminobenzimidazole	Carbendazim	[19]
Carbendazim	Benomyl	[20]
Methyl isothiocyanate	Diphenamid	[21]
Desethylatrazine	Atrazine	[22]

with no atrazine exposure [22]. This is caused by an increase in numbers and activities of DEA degrading bacteria and fungi. Decreased persistence of these transformation products in parent compounds-history soils is important in that there will be less available for movement in surface runoffs.

2.3 Photodegradation

Photodegradation is an important transformation process in the atmosphere, and on the soil and water surfaces. The chemistry and environmental stability of each individual photoproduct can be remarkably different from those of the parent compound and other transformation products [23]. Photoproducts are generally less stable to environmental forces and less toxic than the parent compound. However, some phototransformation products show more toxicity than parent compounds [23]. For example, a recent investigation of photodegradation products of four organophosphorus compounds, azinphos-methyl, chlorpyrifos, malathion, and malaoxon showed formation of toxic intermediate transformation products and further degradation of these toxic transformation products into nontoxic transformation products [24]. Phototransformation products can form after evaporation of parent compounds, or enter the air after formation in soil or in water. Photoproducts of trifluralin were detected in the air, and they arose primarily from the photodegradation of trifluralin on the soil surface followed by volatilization [25]. Phototransformation products can be further degraded by other processes. For example, the photoproduct, photothidiazuron, of the cotton defoliant thidiazuron was further degraded biologically into the more water-soluble hydroxyphenylphotothidiazuron [26].

Most photodegradation information comes from photolysis in organic solvents or on glass surfaces and/or plant metabolism studies [27]. Additional research on photodegradation of transformation products in soil and in air is necessary to understand their fate in the environment, although conditions of these two environmental media are relatively complicated and variable. In the case of soil photolysis, the heterogeneity of soil, together with soil properties varying with meteorological conditions, makes photolytic processes difficult to understand. In contrast to solution photolysis, where light is attenuated by solid particles, both absorption and emission profiles of a pesticide are modified through interaction with soil components such as adsorption to clay minerals or solubilization to humic substances. Extensive investigations of meteorological effects on soil moisture and temperature, as well as development of an elaborate testing chamber controlling these factors, seem to provide better conditions for researchers to examine the photodegradation of pesticides on soil under conditions similar to the real environment [27].

2.4 Movement

Chemicals move away from their initial source sites by various processes in the environment. They can initially travel only short distances by diffusion

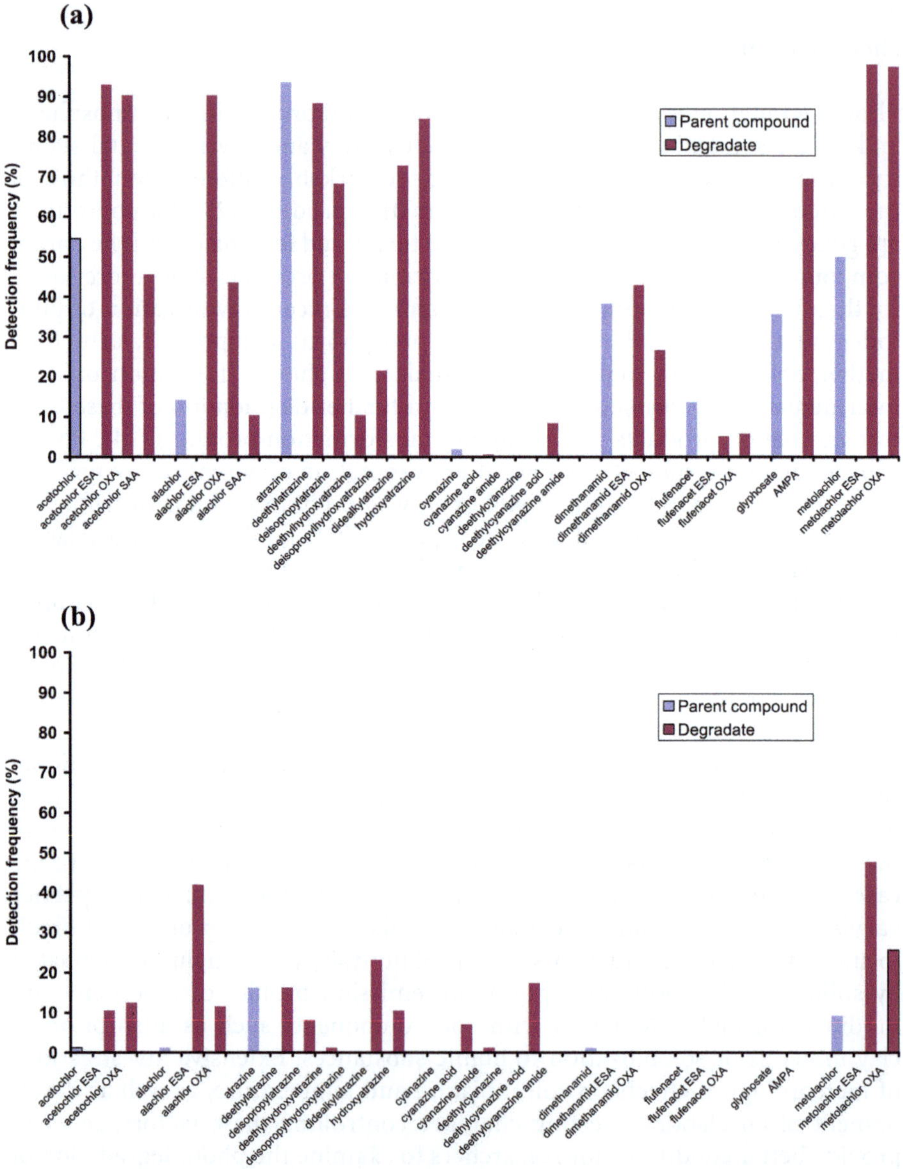

Fig. 4 Detection frequency of some pesticides and their transformation products in the surface water (**a**) and in the ground water (**b**)

into water and air, but can travel longer distances by atmospheric movement, river or ocean currents, and migrating animals. Synthetic chemicals and transformation products have been detected all over the globe in surface water, groundwater, precipitation, air, sediment, and obviously soil. Studies of streams and wells in the Midwestern United States documented the presence of many herbicides and their numerous transformation products, and antibiotics [28, 29]. Figure 4 shows that transformation products are detected as often as or more frequently than the parent compound ([4, 30, 31]; see Kolpin et al., in this volume). The concentrations of some transformation products are greater than their parent compounds, and sometimes only transformation products occurred, with the parent compound no longer detected [32]. An explanation is that transformation products, generally, have higher polarity and water solubility than parent compounds, which favors transport of those transformation products with the aqueous flow. Greater persistence might be another explanation for more frequent detection of some transformation products than their parent compound.

2.5 Leaching

Water solubility and adsorption to soil are important in determining a transformation product's tendency to move through the soil profile with infiltrat-

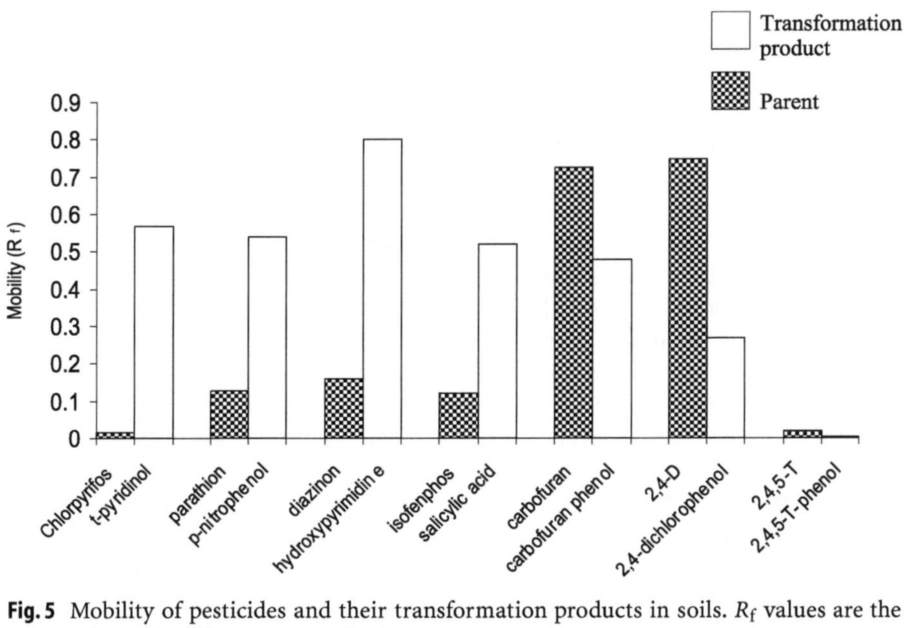

Fig. 5 Mobility of pesticides and their transformation products in soils. R_f values are the average from six different types of soils

ing water. The change of mobility of some transformation products has been confirmed. In one study, the mobilities of pesticides and their transformation products were measured in six different types of soils [32]. Mobility was determined by measuring the R_f value on soil thin-layer chromatography plates, and the results are illustrated in Fig. 5. Parathion, diazinon, isofenphos, and chlorpyrifos were slightly mobile or not mobile in soils, but their transformation products were significantly more mobile than the respective parent compounds. The mobilities of carbofuran phenol and 2,4-dichlorophenol were less than their parent compounds (carbofuran and 2,4-D). The transformation product, 2,4,5-trichlorophenol, was as immobile as its parent compound 2,4,5-T. Another study also showed that hydroxyatrazine was more strongly bound to the soil than atrazine was, and therefore was less likely to leach into ground water [33].

2.6 Volatilization and Atmospheric Transport

Transformation products may get into the air by volatilization, or by wind erosion of solid particles with adsorbed chemicals, or by direct emissions from industrial or agricultural activities. They also can be generated from parent compounds in the atmosphere. Once in the atmosphere they can be transported locally or globally. Some transformation products have greater vapor pressure which is one important factor influencing the volatilization of chemicals. The vapor pressures of PCCH(γ-1,3,4,5,6-pentachlorocyclohexane) and DDE are 14 and 8 times greater than parents compounds lindane and DDT, respectively [34, 35]. Atmospheric chemicals are usually removed by photochemically driven reactions and physical depositional processes such as particles, fog, rain, and snow. The deposited chemicals can reenter the atmosphere by the same mechanism as the original chemicals repeatedly. For example, DDD and DDE, and their parent compound DDT can still be detected in air and precipitation even though the use of DDT has been banned for decades in North America [36]. More than 20 agricultural pesticides have been reported in fog and rainfall in North America and in Europe [37]. Most of these reported pesticides are parent compounds including carbamates, organochlorines, organophosphorus, and herbicides, but also some of their transformation products are also detected. The triazine transformation products, DEA and DIA, were detected in 17.4 and 2.6%, respectively, of the analyzed samples. DEA was detected in more than half of the samples containing atrazine, and trace concentrations of DEA were found in some samples with no detectable atrazine. The ratio of DEA concentrations to atrazine concentrations was greater in rainfall than in streams, indicating atmospheric degradation of atrazine occurs although the mechanism is not yet clear. Mostly, the end products of atmospheric degradation are CO_2, H_2O, sulfate, and nitrate, depending on the chemical composition of the substances that are subject to the degradation processes.

2.7 Binding

Chemicals may attach to soil, vegetation, or other surfaces once they are in the environment. Some portions of chemicals are not recovered by conventional extraction procedures because these bound or unextractable chemical residues are tightly bound to the soil or other solid particles. It is very difficult to differentiate between bound residues of a parent compound and its transformation products. Concerns about bound pesticide residues in soil have been investigated since the late 1960s [33]. The exact mechanism of soil-binding and the ultimate fate of bound residues and their biological significance are not well understood for most chemicals.

Most soil-bound pesticides are less likely to volatilize or to leach through the soil. They are also less easily taken up by plants. In terms of environmental fate, bound residues can be significant and may result in the underestimation of chemical half-life. A nine-year outdoor study showed that soil residues contained as high as 50–60% of the initial applied radioactivity of ^{14}C-atrazine [38, 39], and some transformation products of atrazine persisted for 9 years in outdoor conditions [39]. Most of the bound residues were the hydroxy analogues of atrazine and their dealkylated products.

2.8 Bioavailability and Bioaccumulation

The lower the bioavailability of a toxicant, the less likely it is to have a toxic effect on an organism. This is discussed in more detail in the work by Sinclair and Boxall and Escher et al., in this volume. The availability of a compound for biodegradation is influenced by the compound's location relative to microorganisms and its water solubility. Once absorbed by an organism, the chemical may be metabolized and excreted, sequestered internally, or accumulated. For example, in order to assess toxicity and bioaccumulation of TNT reduction products, 2-amino-4,6-dinitrotoluene (2-ADNT), 4-amino-2,6-dinitrotoluene (4-ADNT), 2,4-diamino-6-nitrotoluene (2,4-DANT), and 2,6-diamino-4-nitrotoluene (2,6-DANT) were tested separately in adult earthworms at different concentrations in a 14-d exposure to amended sandy loam forest soil [40]. TNT, 4-ADNT, and 2-ADNT were lethal to earthworms with a toxicity ranking of 4-ADNT > TNT > 2-ADNT, while 2,4-DANT and 2,6-DANT caused no mortality at tested concentrations. Only TNT transformation products were detected in earthworms, and their 14-d bioaccumulation factors were 5.1, 6.4, 5.1, and 3.2 for 2-ADNT, 4-ADNT, 2,4-DANT, and 2,6-DANT, respectively [40].

Limited bioavailability may lead to unexpected persistence of transformation products in soil and sediment, and longer persistence could lead to accumulation of residues. Whether bound pesticide residues in soils are occluded or may remain bioavailable in the long term in the environment is still an ongoing debate [33]. Generally, soil-bound chemicals are not considered bioavailable prior to desorption [41]. However, some evidence suggests that bound residues can be bioavailable or at least that desorption is not a requisite for biodegradation. Bioavailability is considerably lower from bound residues than from freshly treated soil. It has been suggested that the uptake ratio of chemicals and their transformation products from bound residues compared to those from freshly treated soils was about 1 : 5 [42].

3
Factors Influencing Environmental Fate

The possible environmental fate of chemical transformation products is dependent on various factors. Once formed under certain conditions, transformation products are distributed between four major environmental media: air, water/sediment, soil, and biota. Therefore, environmental fate of transformation products is primarily influenced by the properties of the chemical and conditions of environmental media. Differences in fate are clearly due to variance in environmental media with different physical, chemical, and biological properties including temperature, soil type, light intensity, organic matter, moisture, pH, aeration, and microbial activity. For example, mobilities of five atrazine transformation products, deethylatrazine, deisopropylatrazine, didealkyatrazine, hydroxyatrazine, and ammeline, were negatively correlated with soil organic matter content and positively correlated with sand content [43].

The difference between environmental fate of transformation products compared to parent compounds is mostly due to the different properties of transformation products formed. These important properties include half-life, soil sorption coefficient, water solubility, and vapor pressure. Table 2 lists some changes for the properties of atrazine transformation products [2, 9, 36, 44]. Generally speaking, persistent transformation products are often subject to long-range transport, which might lead to widespread contamination. Water-soluble transformation products more easily run off with rainwater, or leach through the soil as a potential groundwater contaminant. If a transformation product has high soil sorption coefficient, it usually tends to bind to soil and also settle to the sediment in a water system. Transformation products with high vapor pressure are more likely to evaporate into the atmosphere.

Chiral transformation products can be formed from chiral or achiral parent compounds in the environment. Enantiomers are generally considered

Table 2 Properties of atrazine and its selected transformation products

Compounds	Water solubility (mg/l)	K_{OC} (l/kg)	DT50 (d)	Log K_{OW}	Vapor pressure (mPa)
Atrazine	30	129	41–231	2.2–2.8	0.04
Desethylatrazine	3200	56	19–186	1.5	12.44
Desisopropylatrazine	670	61	32–173	1.1–1.2	
Desethyldesisopropylatrazine	600	54	14–68	0.32	
Hydroxyatrazine	5.9	793	32–188	1.4	
Hydroxydesethylatrazine	26.7	927	0.05–7	0.2	
Hydroxydesisopropylatrazine	22	600	0.05-7	- 0.1	
Cyanuric acid	5000	124	6–51		

to have identical physiochemical properties, therefore, their abiotic environmental processes have no difference for both stereoisomers [45]. However, biodegradation processes can be different based on their different biological or toxicological effects. Overall, enantiomers might have different persistence due to chirality in the environment. For example, transformation products of PCBs can be chiral such as hydroxylated PCBs and PCB methyl sulfones. The environmental chemistry and toxicity of hydroxylated PCBs and PCB methyl sulfones have been recently reviewed [46].

Laboratory investigation of transformation products in the environment is very costly, time consuming and often incomplete. Recently, several computer software packages have been developed to predict biotransformation and nonbiological transformation such as photochemical and chemical degradation; some of these approaches have already been described by Wackett and Ellis, in this volume. For example, a tissue metabolism simulator (TIMES) utilizes a heuristic algorithm to generate plausible metabolic maps from a comprehensive library of biotransformation and abiotic reactions, and estimates for system-specific transformation probabilities [47]. Expert judgment is allowed to integrate in TIMES to automatically invoke hydrolysis and many other abiotic reactions, which complete many metabolic pathways. While computer software can suggest possibilities by incorporating various influencing factors, they will not completely replace laboratory experiments.

4 Mass Balance and Environmental Fate

Mass balance-type studies represent an important laboratory experiment, in which the fate of the entirety of a parent compound is modeled in an environmental system, such as a microcosm. Use of a radiolabeled parent compound makes this type of experiment more feasible. Unfortunately, the cost of obtaining and working with a radiolabeled chemical often drives up the expense of these studies. However, the data obtained from such studies are invaluable. The types of radiolabels often used in environmental fate studies of contaminants include ^{14}C, ^{35}S, ^{36}Cl, and ^{3}H, to name a few.

To study the fate of a contaminant in its entirety, i.e. including mineralization, volatilization, partitioning, ^{14}C labeling is often quite useful. For example, uniform ^{14}C labeling of the three carbons in the triazine ring of atrazine with ^{14}C, allows for quantification of mineralization, the degradation of the ring to CO_2 and NH_3 (Fig. 2). This could occur through a soil incubation study that includes traps for CO_2 containing NaOH. Use of a radiolabel also allows for quantification of bound residues, via combustion of soil, sediment, or biotic tissue (e.g. plant material) after those matrices have been extracted. Analysis of solvent extracts of soil, sediment, or tissues at the conclusion of a study timepoint can lead to the identification and quantification of key transformation products. Methods for such analysis could involve the use of high-performance liquid chromatography (HPLC) with a radiodetector. Finally, combustion of extracted soils, sediments, or tissues will account for any bound residues not recovered during the extraction process; a sample oxidizer is often used to accomplish this objective. Bound residues could potentially serve as a source of parent contaminant or transformation products following desorption, from sediment for example, at a later date.

The benefits of a mass balance approach to environmental fate include the ability to trace all aspects of the fate of a contaminant in the environment, including volatilization, uptake, biological or chemical degradation, mineralization, and binding, as outlined in Fig. 1. A mass balance approach also allows the opportunity to identify metabolites using radiodetection. Identification of transformation products using HPLC with UV detection may not yield precise information regarding the chemical structure of the transformation product; however, that data coupled with radiodetection and knowledge of the position of the radiolabel in the parent contaminant can provide evidence for elucidating the identification of a transformation product. A few examples of environmental fate experiments employing a mass balance approach include Henderson and colleagues [47, 48] and Orchard and colleagues [49].

5 Conclusion

The release of synthetic chemicals in the environment may be followed by a very complex series of processes that can transport the chemical and transformation products through air or water, into the ground or even into living organisms. Transformation products of polycyclic aromatic hydrocarbons, nonionic surfactants, musk fragrances, fluorinated alkanes, and polybrominated flame retardants may be of concern in the environment [3]. There are significant gaps in the knowledge of environmental fate of chemical transformation products. Investigation of fate of transformation products is more difficult than parent compounds for a number of reasons. Transformation products are generally present at much lower concentrations in the environment, and often they require more extensive and rigorous process and cleanup prior to reliable analysis by chemical methods. Analytical standards are usually not commercially available for transformation products, and immunochemical methods often show cross-reactivity with metabolites, but at different intensities, which can produce confusing results. Interaction of transformation products with their parent compounds or other substances make it more difficult to investigate their possible fates in the environment.

Generally, transformation products and parent compounds coexist in the environment. They may follow the same fate processes, but they can also have their own unique processes due to the different physiochemical properties of parent compounds and transformation products. With applications of intricate appropriate analytical instrumentation, new sampling or preparation methods, synthesized analytical standards, and new separation techniques, more and more about the environmental fate and effects of transformation products of synthetic chemicals will be understood.

References

1. Corsolini S, Kannan K, Imagawa T, Focardi S, Giesy JP (2002) Polychloronaphthalenes and other dioxin-like compounds in Arctic and Antarctic marine food webs. Environ Sci Technol 36(16):3490–3496
2. Fenner K, Scheringer M, Hungerbuhler K (2003) Joint persistence of transformation products in chemicals assessment: case studies and uncertainty analysis. Risk Anal 23(1):35–53
3. Boxall ABA, Sinclair CJ, Fenner K, Kolpin D, Maud SJ (2004) When synthetic chemicals degrade in the environment. Environ Sci Technol 38(19):368a–375a
4. Halling-Sorensen B, Sengelov G, Tjornelund J (2002) Toxicity of tetracyclines and tetracycline degradation products to environmentally relevant bacteria, including selected tetracycline-resistant bacteria. Arch Environ Contam Toxicol 42(3):263–271
5. Potter TL, Wauchope RD, Culbreath AK (2001) Accumulation and decay of chlorothalonil and selected metabolites in surface soil following foliar application to peanuts. Environ Sci Technol 35(13):2634–2639
6. Tessier DM, Clark JM (1995) Quantitative assessment of the mutagenic potential of environmental degradative products of alachlor. J Agric Food Chem 43(9):2504–2512
7. Somasundaram L, Coats JR (eds) (1991) Pesticide Transformation Products: Fate and Significance in the Environment. Pesticide transformation products in the environment. American Chemical Society, Washington, DC, pp 2–9
8. Matucha M, Gryndler M, Forczek ST, Uhlíová H, Fuksová K, Schröder P (2003) Chloroacetic acids in environmental processes. Environ Chem Lett 1:127–130

9. Kruger EL, Somasundaram L, Kanwar RS, Coats JR (1993) Persistence and degradation of [C-14] atrazine and [C-14] deisopropylatrazine as affected by soil depth and moisture conditions. Environ Toxicol Chem 12(11):1959–1967
10. Somasundaram L, Coats JR (1989) Enhance Biodegradation of Pesticides in the Environment. In: Racke AD, Coats JR (eds) Influence of pesticide metabolites on the development of enhanced biodegradation. American Chemical Society, Washington DC
11. Coats JR (1993) What happens to degradable pesticides? Chemtech 23:25–29
12. Fortin PD, MacPherson I, Neau DB, Bolin JT, Eltis LD (2005) Directed evolution of a ring-cleaving dioxygenase for polychlorinated biphenyl degradation. J Biol Chem 280(51):42307–42314
13. Rajagopal BS, Panda S, Sethunathan N (1986) Accelerated degradation of carbaryl and carbofuran in a flooded soil pretreated with hydrolysis products, 1-naphthol and carbofuran phenol. Bull Environ Contam Toxicol 36(6):827–32
14. Ferris IG, Lichtenstein EP (1980) Interactions between agricultural chemicals and soil microflora and their effects on the degradation of [14C]parathion in a cranberry soil. J Agric Food Chem 28(5):1011–9
15. Racke KR, Coats JR (1988) Comparative degradation of organophosphorus insecticides in soil: specificity of enhanced microbial degradation. J Agric Food Chem 36:193–196
16. Racke KR, Coats JR (1987) Enhanced degradation of isofenphos by soil microorganisms. J Agric Food Chem 35:94–99
17. Somasundaram L, Coats JR (1991) Pesticide Transformation Products: Fate and Significance in the Environment. In: Somasundaram L, Coats JR (eds) Pesticide transformation products in the environment. American Chemical Society, Washington, DC, pp 2–9
18. Avidov E, Aharonson N, Katan J (1990) Involvement of soil microorganisms in the accelerated degradation of diphenamid. Weed Sci 38:186–193
19. Smelt JH, Crum SH, Teunissen W (1989) Accelerated degradation of methyl isothiocyanate in soil. J Environ Sci Health B24:437–455
20. Aharonson N, Katan J, Avidov E, Yarden O (1990) The role of fungi and bacteria in the enhanced degradation of the fungicide carbendazim and the herbicide diphenamid. In: Racke AD, Coats JR (eds) Enhanced Biodegradation of Pesticides in the Environment. American Chemical Society, Washington DC, pp 23–36
21. Roeth FW, Wilson RG, Martin AR, Shea PJ (1990) Enhanced carbamothioate herbicide degradation: research in Nebraska. In: Racke AD, Coats JR (eds) Enhanced Biodegradation of Pesticides in the Environment. American Chemical Society, Washington DC, pp 23–36
22. Arthur EL, Anhalt JC, Anderson TA, Coats JR (1997) Enhanced degradation of deethylatrazine in an atrazine-history soil of Iowa. J Environ Sci Health B 32(5):599–620
23. Miller GC, Crosby DG (1982) Pesticide Photoproducts – Generation and Significance. J Toxicol Clin Toxicol 19(6/7):707–735
24. Bavcon Kralj M, Franko M, Trebse P (2006) Photodegradation of organophosphorus insecticides – Investigations of products and their toxicity using gas chromatography-mass spectrometry and AChE-thermal lens spectrometric bioassay. Chemosphere 67(1):99–107
25. Soderquist CJ, Crosby DG, Moilanen KW, Seiber JN, Woodrow JE (1975) Occurrence of trifluralin and its photoproducts in air. J Agric Food Chem 23(2):304–309
26. Benezet HJ, Knowles CO (1982) Microbial-degradation of thidiazuron and its photoproduct. Arch Environ Contam Toxicol 11(1):107–110

27. Katagi T (2004) Photodegradation of pesticides on plant and soil surfaces. Rev Environ Contam Toxicol 182:1–189
28. Scribner EA, Battaglin WA, Dietze JE, Thurman EM (2003) Reconnaissance data for glyphosate, other selected herbicides, their degradation products, and antibiotics in 51 streams in nine midwestern states, 03–217. US Geological Survey, Reston, VA
29. Kolpin DW, Schnoebelen DJ, Thurman EM (2004) Transformation products provide insight to spatial and temporal trends of herbicides in ground water. Ground Water 42(4):601–608
30. Kolpin DW, Thurman EM, Linhart SM (2001) Occurrence of cyanazine compounds in groundwater: Degradates more prevalent than the parent compound. Environ Sci Technol 35(6):1217–1222
31. Boyd RA (2000) Herbicides and herbicide degradates in shallow groundwater and the Cedar River near a municipal well field, Cedar Rapids, Iowa. Sci Total Environ 248(2/3):241–253
32. Somasundaram L, Coats JR, Racke KD, Shanbhag VM (1991) Mobility of pesticides and their hydrolysis metabolites in soil. Environ Toxicol Chem 10(2):185–194
33. Winkelmann DA, Klaine SJ (1991) Atrazine metabolite behavior in soil-core microcosm: formation, disappearance, and bound residues. Pesticide transformation products in the environment. In: Somasundaram L, Coats JR (eds) Pesticide Transformation Products: Fate and Significance in the Environment. American Chemical Society, Washington DC, pp 75–92
34. Spencer WF, Cliath MM (1970) Vapor density and apparent vapor pressure of lindane (Gamma Bhc). J Agric Food Chem 18(3):529
35. Shen L, Wania F, Lei YD, Teixeira C, Muir DCG, Bidleman TF (2005) Atmospheric distribution and long-range transport behavior of organochlorine pesticides in north America. Environ Sci Technol 39(2):409–420
36. Goolsby DA, Thurman EM, Pomes ML, Meyer MT, Battaglin WA (1997) Herbicides and their metabolites in rainfall: Origin, transport, and deposition patterns across the midwestern and northeastern United States, 1990–1991. Environ Sci Technol 31(5):1325–1333
37. Gevao B, Mordaunt C, Semple KT, Piearce TG, Jones KC (2001) Bioavailability of nonextractable (bound) pesticide residues to earthworms. Environ Sci Technol 35(3):501–507
38. Capriel P, Haisch A, Khan SU (1985) Distribution and nature of bound (nonextractable) residues of atrazine in a mineral soil 9 years after the herbicide application. J Agric Food Chem 33(4):567–569
39. Lachance B, Renoux AY, Sarrazin M, Hawari J, Sunahara GI (2004) Toxicity and bioaccumulation of reduced TNT metabolites in the earthworm *Eisenia andrei* exposed to amended forest soil. Chemosphere 55(10):1339–1348
40. Park JH, Feng YC, Ji PS, Voice TC, Boyd SA (2003) Assessment of bioavailability of soil-sorbed atrazine. Appl Environ Microbiol 69(6):3288–3298
41. Khan SU (1982) Bound pesticide-residues in soil and plants. Residue Rev 84:1–25
42. Kruger EL, Zhu BL, Coats JR (1996) Relative mobilities of atrazine, five atrazine degradates, metolachlor, and simazine in soils of Iowa. Environ Toxicol Chem 15(5):691–695
43. Prosen H, Zupančič-Krajl L (2005) Evaluation of photolysis and hydrolysis of atrazine and its first degradation products in the presence of humic acids. Environ Pollut 133(3):517–529
44. Muller TA, Kohler HPE (2004) Chirality of pollutants – effects on metabolism and fate. Appl Microbiol Biotechnol 64(3):300–316

45. Letcher RJ, Klasson-Wehler E, Bergman A (2000) Methyl sulfone and hydroxylated metabolites of polychlorinated biphenyls. Handb Environ Chem 3:315–359
46. Mekenyan OG, Dimitrov SD, Pavlov TS, Veith GD (2004) A systematic approach to simulating metabolism in computational toxicology. I. The TIMES heuristic modelling framework. Curr Pharmaceut Des 10(11):1273–1293
47. Henderson KL, Belden JB, Coatst JR (2007) Fate of atrazine in a grassed phytoremediation system. Environ Toxicol Chem 26(9):1836–1842
48. Henderson KL, Belden JB, Coats JR (2007) Mass balance of metolachlor in a grassed phytoremediation system. Environ Sci Technol 41(11):4084–4089
49. Orchard BJ, Doucette WJ, Chard JK, Bugbee B (2000) A novel laboratory system for determining fate of volatile organic compounds in planted systems. Environ Toxicol Chem 19(4):888–894

Hdb Env Chem Vol. 2, Part P (2009): 121–149
DOI 10.1007/698_2_013

Published online: 6 March 2008

Modelling Environmental Exposure to Transformation Products of Organic Chemicals

Kathrin Fenner[1,2] (✉) · Urs Schenker[3] · Martin Scheringer[3]

[1]Swiss Federal Institute for Aquatic Science and Technology (Eawag), PO Box 611, 8600 Dübendorf, Switzerland

[2]Institute of Biogeochemistry and Pollutant Dynamics, Swiss Federal Institute of Technology (ETH), ETH Zürich, 8092 Zurich, Switzerland
kathrin.fenner@eawag.ch

[3]Institute for Chemical and Bioengineering, Swiss Federal Institute of Technology (ETH), ETH Zürich, 8093 Zurich, Switzerland

Abstract Transformation products of environmental contaminants are likely to contribute significantly to the overall chemical pressure on valuable environmental resources. However, the whole extent of this so far often over-looked additional contamination remains unclear because the number of monitoring studies addressing transformation products is currently small and they are largely focused on well-known pesticide transformation products. Environmental fate modelling of transformation products opens up the possibility to predict the likely presence of environmental transformation products in environmental compartments of interest and to point towards new, potentially relevant transformation products. Whereas, depending on their purpose, there are different types of fate models for transformation products, this chapter will focus on multispecies multimedia models because of their general applicability to various tasks in chemical risk assessment and quality assessment of environmental resources. The chapter will introduce the mathematical framework underlying most multispecies multimedia models and discuss its application to three examples. These examples include the extension of overall persistence to include transformation products, the assessment of the long-range transport potential of persistent transformation products of semivolatile organic compounds, and the prediction of relative concentrations of pesticide transformation products in surface water bodies. The second part of the chapter discusses data requirements and availability for multispecies multimedia models and sheds some light on the accuracy of frequently used chemical property estimation tools. Lastly, tools to predict transformation schemes in those cases where no information on possible transformation products is available are introduced and their current limitations are discussed.

Keywords Concentration prediction · Joint persistence · Multispecies multimedia models · Pesticides · Structure–property estimation methods

Abbreviations

BOD	Biological oxygen demand
DDD	*p,p′*-Dichlorodiphenyldichloroethane [72-54-8]
DDE	*p,p′*-Dichlorodiphenyldichloroethylene [72-55-9]
DDT	*p,p′*-Dichlorodiphenyltrichloroethane [50-29-3]
eACP-10	Arctic contamination potential relative to total emission after 10 years
ff	Fraction of formation
JP	Joint persistence
K_H	Henry's law constant
K_{oc}	Organic carbon-water partition coefficient
K_{ow}	Octanol-water partition coefficient
LRTP	Long-range transport potential
OECD	Organisation for Economic Co-operation and Development
pp-LFER	Polyparameter linear free-energy relationship
P_{ov}	Overall persistence
PSM	Primary survey model
QSAR	Quantitative structure-activity relationship
RAC	Relative aquatic concentration
REACH	Registration, evaluation and authorization of chemicals
SSR	Secondary spatial range
UM-PPS	University of Minnesota pathway prediction system

1 Introduction

Environmental fate models are widely used to predict concentrations of chemicals in different environmental compartments. Their predictions are based on a representation of the environmental conditions and processes considered relevant and the physico-chemical properties and environmental half-lives of the chemicals in question. Environmental compartments of particular interest include those that might lead to human exposure such as surface and groundwater abstracted for drinking-water production, and those representing valuable ecosystems such as surface water bodies and diverse soils. The goal of environmental fate modelling of transformation products is to predict the likely presence of environmental transformation products in the compartments of interest. Here the term transformation products is meant to cover all products formed in the environment from parent compounds that have been purposely or accidentally released into the environment. Transformation products might be formed by abiotic or biotic processes such as hydrolysis, photolysis, or biodegradation by bacteria and fungi. We distinguish transformation products from metabolites formed in mammalian metabolism, which are not covered by the models presented here, but are usually dealt with in physiologically based pharmacokinetic/pharmacodynamic (PBPK/PD) modelling (see for example [1]).

While for a long time environmental fate and exposure modelling was focused on predicting the fate of parent compounds of interest, the last decade has seen the development of a number of models that account for transformation product formation and fate. Depending on their specific context and purpose, these models differ widely with regard to how environmental fate processes are described and parameterized, and hence with regard to their complexity and flexibility to be applied to different compounds. However, most of them can be roughly attributed to one of the three following classes of models: (i) methods for ranking large sets of transformation products with regard to their risk, (ii) multispecies multimedia models, used for environmental risk and quality assessment, e.g., for registration purposes, (iii) simulation models that predict site-specific environmental concentrations and are mainly used to investigate the processes governing the fate of specific transformation products.

After a short review of the three classes of fate models for transformation products, the chapter will focus on multispecies multimedia models because of their general applicability to various tasks in chemical risk assessment and quality assessment of aquatic and terrestrial resources. Moreover, as will be detailed later, the data situation for transformation products, regarding both monitoring as well as chemical property data, is often such that the use of more complex, data-intensive models is neither warranted nor would it improve the accuracy of the predictions. The chapter will introduce the mathematical framework underlying most multispecies multimedia models and discuss its application to three examples for different assessment endpoints. These examples include the extension of overall persistence to include transformation products, the assessment of the long-range transport potential of persistent transformation products of semivolatile organic compounds (SOCs), and the prediction of relative concentrations of pesticide transformation products in surface water bodies. The second part of the chapter will discuss data requirements and availability for multispecies multimedia models and shed some light on the accuracy of some frequently used chemical property estimation tools. Finally, tools to predict transformation schemes in those cases where no information on possible transformation products is available are introduced and their current limitations are discussed.

2 Environmental Fate Models for Transformation Products: An Overview

2.1 Ranking Methods

Knowledge of the physico-chemical properties, degradability, and extent of formation of transformation products is generally very scarce. Ranking or prioritization methods aim to make fullest possible use of the available data to identify transformation products that might pose a potential risk to human or environmental health. The intention is that these rankings might then serve as a starting point for more in-depth studies on the highest-ranked transformation products. These might include the generation of experimental fate and toxicity information, analytical method development, or the planning of targeted monitoring studies for these compounds. Belfroid et al. [2] developed a method to assess the risk of pesticide transformation products to aquatic ecosystems relative to their parent pesticide. Order-of-magnitude changes in physico-chemical parameters, in compartmental half-lives or in ecotoxicological endpoints are used to assess whether risk is enhanced or reduced in comparison to the parent pesticide. The analysis was carried out for 20 regularly used pesticides and 78 of their transformation products. Triazines, carbamates, and phenoxypropionic acids were tentatively identified as showing a tendency for the formation of hazardous transformation products. Sinclair et al. [3] developed a risk-based scoring method for identifying the most important transformation products in drinking water resources. Indices for usage of the parent pesticide, and mobility, persistence and human health effects of the transformation products between 0 and 1 are determined and combined into a final risk score. The method was applied to 122 pesticides and their 371 transformation products in Great Britain and to 33 pesticides and their 86 transformation products in California. The transformation products were grouped according to different degrees of data availability and, within each group, compounds with a high risk index were identified. For compound classes other than pesticides, no such methods to rank transformation products have been reported, though it would be relatively easy to refine some of these current methods for selected chemical classes if the transformation products are known.

2.2 Multispecies Multimedia Models

Multispecies multimedia models are used to calculate the distribution and fate of parent compounds and transformation products in a multi-compartment environment, including formation and interconversion of transformation products in each compartment. These processes are expressed through a set of coupled mass-balance equations that are solved simultaneously. Fenner et al. [4] first proposed the general mathematical framework for modelling the fate of any number of transformation products in a generic multimedia environment and employed it to define and calculate joint persistence (JP), a measure of the overall persistence of a parent compound and its transformation products [5, 6]. They used the same mathematical framework but a more region-specific model to carry out a risk assessment for nonylphenol polyethoxylates in combination with their transformation products in Swiss surface waters [7] and, more recently, to predict relative aquatic concentrations (RAC) for a set of 16 commonly used pesticides and 53 transformation products in a generic river water body [8]. Using a similar approach, Quartier et al. [9] defined the secondary spatial range (SSR) of transformation products as a measure of long-range transport potential for transformation products. They derived an analytical solution to calculate SSR in a multimedia environment, assuming instantaneous equilibrium between the compartments. Based on the fugacity concept, Cahill et al. [10] later also developed a multispecies multimedia modelling framework which, however, remained restricted to up to four interconverting chemical species because the equations are solved in a more calculation-time intensive manner. The model was further developed into a more region-specific, high-resolution model that was applied to simulate the fate of malathion and pentachlorophenol and their transformation products [11]. Although the authors demonstrated that (at least for malathion) more accurate results are obtained with the high-resolution model, they also acknowledged that the strongly enhanced data needs of this type of model are unrealistic for most transformation product problems and will often lead to additional parameter uncertainty, which may override the gain in model accuracy. More recent developments in this class of models include the extension of the general framework to cover formation of transformation products in aquatic food webs and their biomagnification [12] as well as the extension of the spatially resolved, global multimedia model CliMoChem to calculate the long-range transport potential of transformation products of semivolatile organic chemicals [13].

2.3 Site-Specific Simulation Models

Simulation models are mostly site-specific, i.e., they are used to predict realistic concentrations at a given site. In conjunction with field data, such models can help to gain insight into the processes governing chemical fate at the field or regional scale. They are therefore mostly considered research tools and are only occasionally also used in an assessment context, e.g., in higher tiers of pesticide risk assessment in the EU where process-based soil simulation models are used to calculate time-resolved runoff and leachate concentrations at the field scale [14]. Due to the complexity and data intensity of simulation models, transformation products are mostly treated as separate species in these models, i.e., their formation is not dynamically coupled to the fate of the parent compound (see, for example [15, 16]). More recently, however, coupling of the fate of parent compound and transformation products has also been sought in simulation models. One such example is the modelling of nonylphenol ethoxylates and their transformation products in a Dutch estuary [17]. There, the simulation results were used in combination with field data to derive aqueous biodegradation half-lives for all species involved. Another example is the extension of the root zone water quality model (RZWQM) for the prediction of runoff so that two transformation products can be handled in parallel or in series. This model was evaluated against 2 years of field data from a mesoplot rainfall-runoff simulation experiment for fenamiphos, its first-generation transformation product fenamiphos sulfoxide and its second-generation transformation product fenamiphos sulfone [18].

3 Multispecies Multimedia Models: Mathematical Framework and Applications

3.1 General Model Structure

In multimedia box models, the environmental fate of a chemical is described by a set of coupled mass-balance equations for all boxes of the model. These equations include terms for degradation, inter-media exchange such as settling and resuspension of particles, and transport with air and water flows [19, 20]. Equations for different boxes are coupled by inter-media exchange terms (linking different environmental media) and terms for transport with air or water (linking spatial segments of the model). These processes are assumed to be first-order processes so that the entire system can be described by a set of linear ordinary differential equations.

An efficient way to treat such a system is to assemble all coefficients of the different terms of the mass-balance equations in a matrix and to apply methods of matrix algebra to solve the system for steady-state concentrations (level III) or for the concentrations as functions of time (level IV) [19]. We denote the matrix of coefficients (the "fate matrix") by $\boldsymbol{S}$, the vector of concentrations in all boxes of the model by $\boldsymbol{c}$, and the vector of all source terms by $\boldsymbol{q}$. The set of mass-balance equations describing the temporal changes of the concentrations in all boxes then reads: $\dot{\boldsymbol{c}} = -\boldsymbol{S} \cdot \boldsymbol{c} + \boldsymbol{q}$. The steady-state solution is obtained by setting $\dot{\boldsymbol{c}}$ equal to zero and solving for $\boldsymbol{c}$. This leads to $\boldsymbol{c}^{ss} = \boldsymbol{S}^{-1} \cdot \boldsymbol{q}$, i.e., to obtain the steady-state concentrations the emission vector has to be multiplied by the inverse of the matrix S. For the dynamic solutions of the system, the eigenvalues and eigenvectors of $\boldsymbol{S}$ have to be determined.

If a single chemical is considered, the terms in the matrix $\boldsymbol{S}$ simply describe loss and transport processes for that chemical. If transformation products are included, the mass of the parent compound that is degraded does not disappear from the system but is converted completely or in parts into transformation products. To account for this in the mass-balance equations, the degradation term for the parent compound appears as a source term for transformation products (multiplied by a fraction of formation, ff_i, between 0 and 1, indicating the fraction of parent compound that is converted into transformation product i). The same applies to the formation of second- and higher-generation products. For each transformation product i, a full set of mass-balance equations represented by the fate matrix $\boldsymbol{S}_i$ has to be set up in the same way as the parent compound. In other words, if the model considered has n boxes, then there are n equations for the parent compound, and each transformation product adds another n equations to the system. Hence, for a system with a parent compound forming $m - 1$ transformation products, the fate matrix $\boldsymbol{S}$ describing the whole system has a dimension of $nm \times nm$. In the matrix $\boldsymbol{S}$, the fate matrices $\boldsymbol{S}_i$ for each species lie on the diagonal of $\boldsymbol{S}$, whereas the transformation submatrices $\boldsymbol{K}_{ij}$, which include the rate constants of formation of j out of i, lie at the position ij in the matrix $\boldsymbol{S}$ [4, 12]. This is illustrated in Fig. 1, which shows a model matrix for a given reaction scheme with two first-generation transformation products, B and C, and one second-generation transformation product, D.

To set up such a model system, relevant transformation products have to be selected in the first step. In many cases, there are transformation products of a parent compound that are not formed in high amounts or are rapidly transformed into a more persistent transformation product. Such compounds do not have to be included in the transformation scheme. Identification of those transformation products that are formed in significant amounts and are sufficiently persistent requires knowledge of the biotic and abiotic transform-

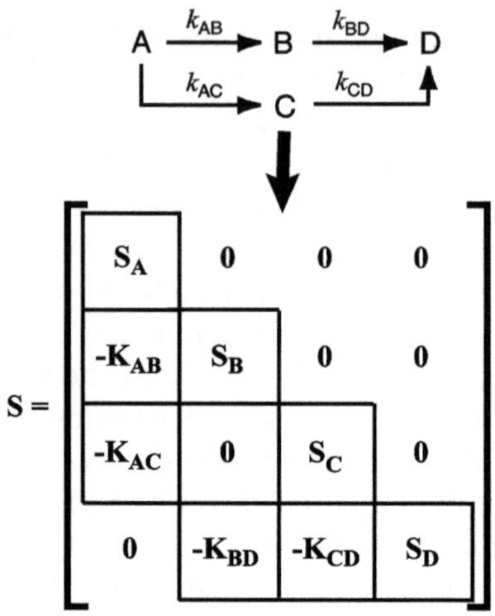

Fig. 1 Construction of the matrix S for an example in which parent compound A decays in two parallel reactions into the transformation products B and C, which are both further transformed to D, hence $m = 4$. The rate constants of the transformation processes are denoted by k_{ij}. $\boldsymbol{S}_i$ is the fate matrix of the species i including its degradation and transfer processes. $\boldsymbol{K}_{ij}$ is the transformation matrix representing the transformation of the species i to the species j. Reprinted with permission from [4], p 3810. © (2000) American Chemical Society

ation pathways of the parent compound. Often, such knowledge is not readily available; in Sect. 4.2 expert systems for investigating transformation schemes and identifying relevant transformation products are discussed.

Additional parameters needed for the mass-balance equations are the physico-chemical properties of all transformation products considered, the degradation rate constants, and the fractions of formation of all transformation reactions. The fractions of formation account for the generation of several transformation products in parallel and for yields of less than 100%. For example, if two products are formed in roughly equal amounts and about 80% of the precursor is known to be converted into these two products, their fractions of formation are 0.4. Fractions of formation can be derived from kinetic information about a transformation pathway (see Sect. 4.1). However, because this information is often missing, most fractions of formation have to be estimated.

3.2 Modelling of Joint Persistence

Overall persistence (P_{ov}) is a measure of the period during which an environment or region is exposed to a chemical of interest. In more technical terms, it is the reactive residence time of a chemical in a closed environmental system consisting of different media, i.e., the ratio of the mass present in the system divided by the mass flux through the system due to degrading reactions, which is equal the total emissions into the system (Eq. 1).

$$P_{\mathrm{ov}} = \frac{\sum_j m_j^{\mathrm{ss}}}{\sum_j q_j}, \tag{1}$$

where m_j^{ss} [mol] is the steady-state mass and q_j [mol/d] is the mass flux into each environmental medium j in the environmental system under consideration. Besides toxicity, bioaccumulation, and long-range transport potential, persistence is one of the crucial criteria in several national and international regulatory frameworks to identify highly hazardous compounds [21]. These frameworks, however, rely on single-media half-life criteria rather than on P_{ov} for the assessment of persistence. This has the disadvantage that chemicals might be categorized as persistent or non-persistent based on half-lives in compartments where only negligible amounts of the compounds reside. Recently, both the scientific and policy communities have pointed out that multimedia models and P_{ov} are valuable tools to assess persistence in a more realistic manner [22].

P_{ov} is commonly calculated based on single-media half-lives that have been derived from degradation simulation studies such as those recommended in OECD 307 [23] and OECD 308 [24]. In these studies, disappearance of the chemical in question is followed as a function of time. The half-lives thus obtained therefore usually do not account for the formation of transformation products, which are likely to prolong exposure of the environment to chemicals that may still exhibit hazards similar to the parent compound. Joint persistence (JP) has therefore been suggested as a new indicator for overall persistence that reflects the prolonged exposure of the environment due to the formation of transformation products [4, 5] (Eq. 2). For better distinction, the overall persistence of the parent compound alone will be termed primary persistence (PP) for the remainder of the chapter.

$$\mathrm{JP} = \frac{M_{\mathrm{PC}}^{\mathrm{ss}} + \sum_{i=1}^{n} M_i^{\mathrm{ss}}}{q_{\mathrm{PC}}} \tag{2}$$

with M_i^{ss} being the steady-state mass (mol) of each compound summed over all compartments j ($M_i^{\mathrm{ss}} = \sum_{j=s,a,w} m_{i,j}^{\mathrm{ss}}$), n the number of TPs, and q_{PC} the release rate (mol/d) of the parent compound.

JP has been calculated for 16 current-use pesticides [8], the surfactant nonylphenol polyethoxylate (NPnEO), and the solvent perchloroethylene [5], based on a thorough compilation of experimental data on compound properties and transformation schemes (if such data are not available, they have to be estimated as described in Sect. 4.2). In Fig. 2, the JP results for the 18 compounds are displayed.

The results clearly show the importance of including transformation products in persistence assessments. For 12 of 18 compounds, the JP more than doubles compared to the PP. However, the ratio *Q* between JP and PP usually does not exceed a value of 4 unless the parent compound is a fast hydrolyzing pro-pesticide (bromoxynil-oct, fluoroglycofen-et and kresoxim-me). Even so, accounting for transformation product formation can have a considerable effect on compound classification. If, for the purpose of illustration and in accordance with the 60 days water half-life criterion in the Stockholm Convention on Persistent Organic Pollutants [25], a value of 60 days is used as persistence criterion, 8 of the 18 compounds that would not be classified as persistent according to their PP, have JP values exceeding 60 days, classifying the entire substance family as persistent (Fig. 2).

It has been argued that if half-life estimates are derived from mineralization rates, the consideration of transformation products in persistence assessment is not necessary. This might be a reasonable first approximation. However, this argument is only partially correct. First, in many cases it is not clear whether reported half-lives actually describe complete mineralization of the parent compound. Second, single-compartment mineralization rates fail to reflect situations where, during the course of mineralization in

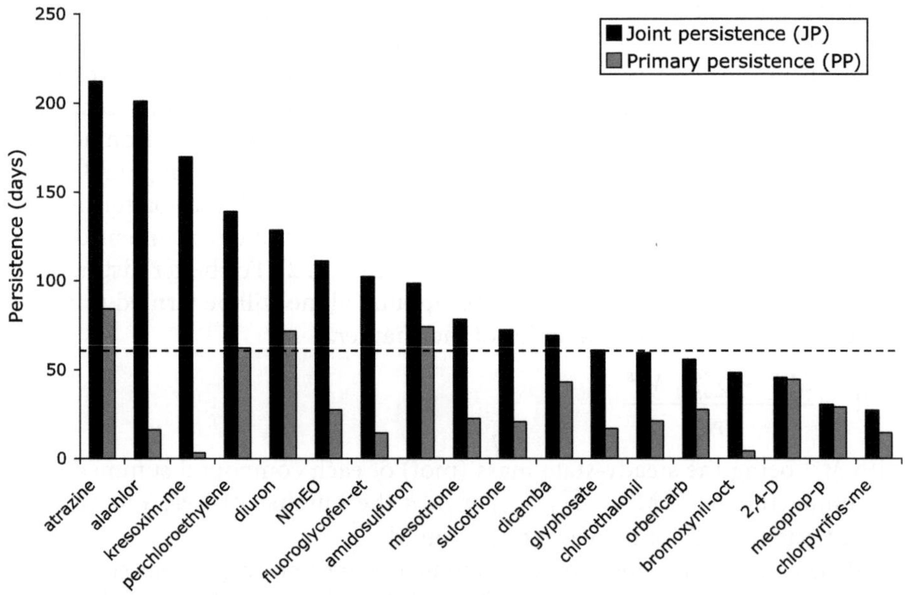

Fig. 2 Primary (PP) and joint persistence (JP) for 16 pesticides, the surfactant nonylphenol polyethoxylate (NPnEO), and the solvent perchloroethylene sorted according to JP. The *dashed line* indicates the 60-day water half-life used as persistence criterion in the Stockholm Convention [25]. Reprinted with permission from [8], p 2447. © (2007) American Chemical Society

one compartment, transformation products are being formed that are efficiently transported into another environmental compartment where they exhibit a distinctly different behavior. One such example is the formation of highly polar pesticide transformation products in soil, which are then efficiently washed out into water bodies where their persistence is usually much higher than in a biologically active soil.

3.3 Modelling of Long-Range Transport Potential of Persistent Transformation Products

3.3.1 Necessity of Geographically Resolved Models

It is generally acknowledged that persistence is not the only requirement for a substance to be present at remote locations [26, 27]. In addition to being recalcitrant to degradation, the substance must also be sufficiently mobile to be transported over long distances; in other words, it must have a high long-range transport potential (LRTP). Therefore, multimedia box models and atmospheric transport models have been developed that simulate substance transport by atmospheric and oceanic transport processes (wind, ocean currents), see the overview in [28]. Multimedia box models are constructed as a series of interconnected zones that each contain the same media that can usually be found in single-box models. Inside a zone, the different media are linked by exchange processes, whereas between zones transport processes take place. Generally, the environmental conditions (temperature, vegetation, organic matter content in soil, OH radical concentration) are set to be different in each box or zone. Established models of this type are GloboPOP [29] and CliMoChem [30, 31] (both zonally averaged); an atmospheric transport model that has also been adjusted for semivolatile organic chemicals is ECHAM4 [32].

The CliMoChem model has recently been adapted to include transformation products [13]: as described for single-box models above, the exchange matrix in CliMoChem has been extended to contain a variable number of sub-

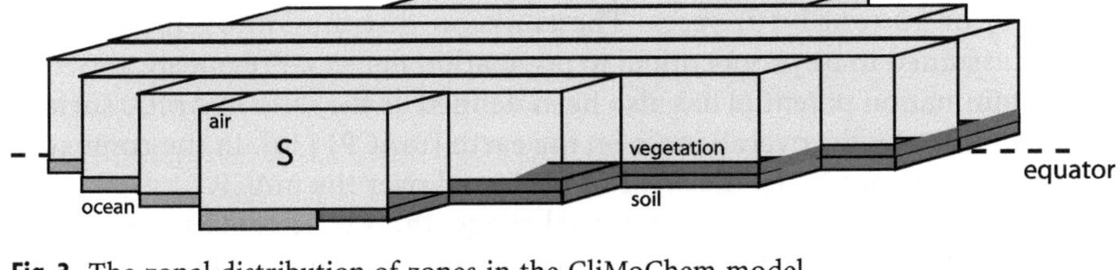

Fig. 3 The zonal distribution of zones in the CliMoChem model

stances that are related to one another through transformation processes. The transformation processes and the relative importance of different transformation pathways may differ between the environmental media (for instance OH reaction in the atmosphere, but (an)aerobic biodegradation in soil).

To better quantify the mobility of substances, indicators that can be calculated from the model results have been developed. Indicators for long-range transport potential are the spatial range [26, 27], the characteristic travel distance (CTD) [33], the Great Lakes Transport Efficiency (GLTE) [34], and the Arctic contamination potential [35]. In the following two sections, we describe how such indicators have been adapted for transformation products.

3.3.2 Spatial Range and Joint Spatial Range

Spatial range is defined as the 95% interquantile range of the geographical distribution of the time-integrated concentration of a compound on a north–south transect of the earth after an emission at the equator [30]. In analogy to the definition of the joint persistence, it is possible to define a joint spatial range [13]: it is calculated from the sum of the time-integrated concentrations of the parent compound and all transformation products, instead of from the time-integrated concentration of the parent compound only. The 95% interquantile range of the different transformation products alone is the apparent spatial range of the transformation products. Note that the apparent spatial range of the transformation products is not equal to the spatial range of those substances if they were emitted as a single pulse at the equator. Rather, the apparent spatial range is obtained after the degradation of a parent compound: therefore, the transformation product is continuously formed in the environment at those places where the parent compound is present and being degraded.

3.3.3 Arctic Contamination Potential and Joint Arctic Contamination Potential

The Arctic contamination potential (ACP) focuses specifically on chemicals that tend to accumulate in Arctic surface media and has therefore been defined as the ratio of the substance mass in Arctic surface media (all media except the atmosphere) divided by the total emissions of substance after 1 (eACP-1) or 10 (eACP-10) years. The emission distribution on the globe is mostly assumed to be proportional to population density. Previously, the Arctic contamination potential has also been defined as the ratio in Arctic surface media divided by the overall mass on the earth (mACP) [36]. In the context of transformation products, the eACP is preferred over the mACP.

In analogy to persistence and spatial range, the Arctic contamination potential can be extended to include transformation products. The joint Arctic contamination potential (joint eACP) has been defined as the ratio of the mass of all the substances in the Arctic surface media divided by the total emissions of the parent compound after 1 or 10 years [13].

3.3.4 Impact of Transformation Products on Spatial Range and Arctic Contamination Potential of Legacy Pesticides

As an illustrative example, the importance of the transformation products of DDT (DDE and DDD), aldrin (dieldrin, a pesticide itself), and heptachlor (heptachlorepoxide) are discussed. The transformation products of these substances have been widely detected and are sometimes present in higher concentrations than their parent compounds.

In the calculations presented, heptachlor is degraded into heptachlorepoxide in all environmental media (with a fraction of formation $ff = 0.9$), aldrin is degraded into dieldrin in all environmental media ($ff = 0.9$), too, whereas DDT degrades into DDE in the atmosphere ($ff = 0.9$), and in equal parts (both $ff = 0.5$) into DDE and DDD in all the other media. Degradation half-lives were extracted as experimental values from the literature [37, 38] where possible, or calculated with QSAR software (especially for OH reactions) [39].

The spatial ranges calculated with the extended CliMoChem model (Fig. 4) indicate that for heptachlor, the joint spatial range is clearly higher than the spatial range of the parent compound alone. For the aldrin and DDT substance families, on the contrary, the joint spatial ranges are similar to the spatial range of the parent compound alone. At first sight, this is unexpected because, as can be seen for the transformation products DDE and DDD, both have apparent spatial ranges greater than that of DDT. Closer examination of the results, however, shows that they both exhibit lower persistence than DDT and therefore do not significantly influence the joint spatial range.

In terms of the Arctic contamination potential, the model correctly predicts that the transformation products of aldrin and heptachlor will be present in much higher concentrations in the Arctic than the parent compound: measurements in the Arctic do indeed predominantly detect dieldrin and heptachlorepoxide, but no or very little aldrin or heptachlor [40]. For the case of DDT, the model predicts that DDT should largely dominate over its transformation products in the Arctic. In reality, DDT is indeed often measured in the highest concentrations, but DDE is usually present in concentrations of similar magnitude. This inconsistency between measurements and modelling results is probably due to the high uncertainties in the atmospheric half-lives of the two substances: whereas for DDT, measured half-lives for reaction in air were available [41], no such measurements exist for DDE and DDD, and therefore half-lives had to be estimated with QSARs [39]. For DDT, the half-lives predicted using QSAR tools are about one order of magni-

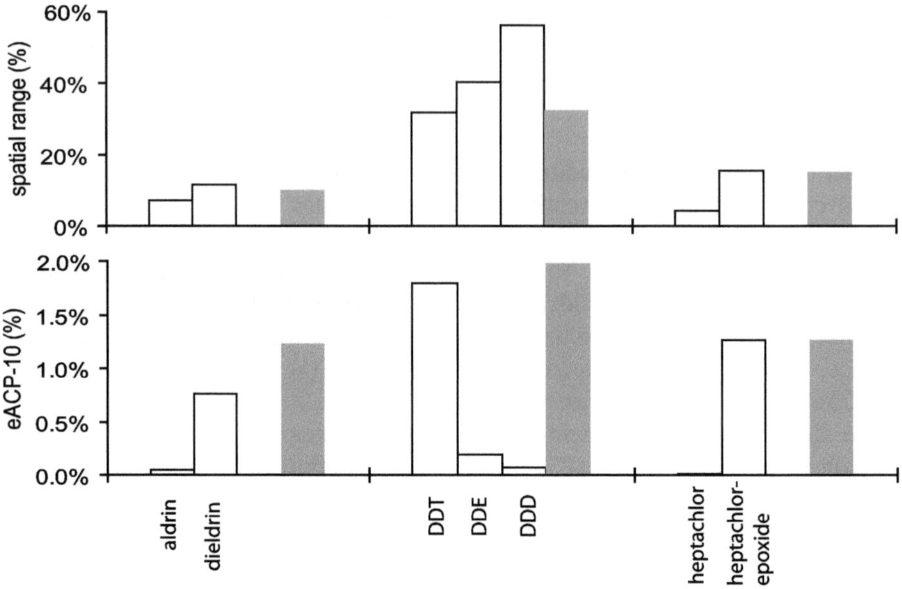

Fig. 4 Spatial range (*top*, in % of the distance between North and South Poles), and Arctic contamination potential (*bottom*) of three legacy pesticides and their first-generation transformation products. In the *top panel*, the *white bars* represent the (apparent) spatial range of the parent compound and the degradation products, whereas the *grey bars* represent the joint spatial range. In analogy, for the *bottom panel*, the *white bars* represent the eACP-10 of the parent compound and the individual transformation products, whereas the *grey bar* represents the joint eACP-10. Modified from [13]. © (2007) ecomed publishers

tude smaller than the experimentally determined values. Given the structural similarities between DDE, DDD, and DDT, this suggests that the QSAR values of DDE and DDD are likely to underestimated the actual values as well.

Finally, there are large differences between the Arctic contamination potential and the spatial range for the DDT transformation products: whereas DDE and DDD have apparent spatial ranges higher than DDT, their eACP-10 is negligible compared to DDT. The definitions of Arctic contamination potential and spatial range explain these differences: the eACP-10 of the transformation products compares the amount of DDE and DDD present in Arctic surface media to the amount of DDT emitted. Since DDE and DDD are present only in small concentrations, their corresponding eACP-10 is low. The apparent spatial range, however, is the 95% interquantile range of the distribution of the transformation products themselves, and is independent of the amount of parent compound emitted. Given the wide-ranging distributions of DDE and DDD, their apparent spatial range is high, even if the absolute amount of the substances is low. The relative importance of the transformation products can only be seen when the joint spatial range is compared to the spatial range of the parent compound. In our case, this difference is small, confirming that the small quantities of DDE and DDD have little influence on the global distribution of the whole substance family.

3.4 Aquatic Concentrations of Transformation Products of Micropollutants

Another important aspect in chemicals assessment, besides the hazard-based P_{OV} and LRTP criteria, is how a chemical and its transformation products affect water quality. Water quality with respect to both human and ecosystem health is mostly assessed in a risk-based manner, i.e., measured or predicted environmental concentrations are compared to toxicity thresholds. When it comes to assessing the role of transformation products for water quality, it is therefore crucial to be able to predict or measure the concentrations of the transformation products relative to each other and relative to the parent compounds.

A long-term monitoring program of the U.S. Geological Survey (USGS) for herbicides (mainly triazines and chloroacetanilides) and their transformation products in U.S. midwestern streams and groundwater reports total transformation product concentrations to be about 20-fold higher than total parent compound concentrations and frequencies of detection for transformation products to be systematically higher than for parent compounds [42–44]. These findings suggest that mobile transformation products might make up a major share of the total exposure to chemical compounds in water resources. They further raise the question of which other transformation products than the well-known triazine and chloroacetanilides transformation products might be present in our water resources, but are currently not being monitored.

Since the number of possible transformation products obviously exceeds the number of chemicals in commerce, carrying out a full-fledged assessment of every possible transformation product is not feasible. Instead, screening approaches are needed that efficiently prioritize transformation products for further investigations according to their aquatic exposure potential and their toxicological and ecotoxicological effects. Such approaches for pesticide transformation products have been described by Belfroid et al. [2] and Sinclair et al. [3]. However, these methods do not (or only to a limited extent) account for the dynamics of transformation products formation and transport because they are based on qualitative or semi-quantitative scores only. For chemicals that are mainly released to soil, such as pesticides and veterinary pharmaceuticals, Gasser et al. [8] developed a process-based model and corresponding indicators that quantitatively reflect the environmental fate of parent compounds and transformation products.

The model consists of an air and a soil compartment that is connected to an average river body through runoff and erosion (Fig. 5). The purpose of this

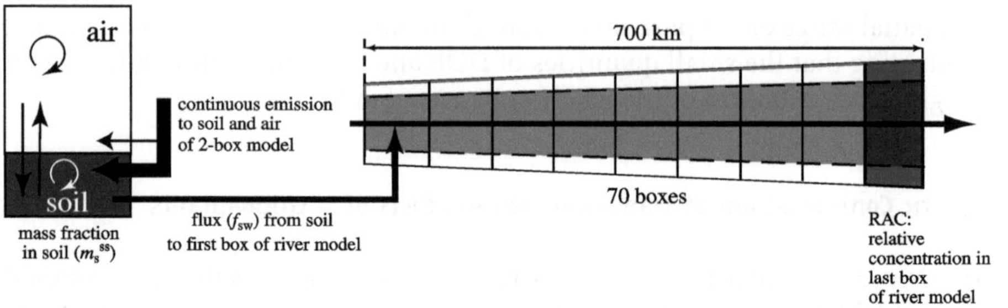

Fig. 5 Scheme of model for calculating relative aquatic concentrations (RAC). The soil compartment and the river model are connected through the flux (f_{SW}) from soil to the first box of the river model. The river model consists of flowing water (*shaded boxes*), stagnant water (*white boxes*) and an underlying river sediment (not shown). Reprinted with permission from [8], p 2446. © (2007) American Chemical Society

model is not to simulate a particular region, but to represent an evaluative or generic region in which all transformation products can be consistently assessed and whose level of detail matches the available substance data. This model setup has been used to calculate relative aquatic concentrations (RAC) for a set of transformation products and their parent compounds for unit emission rates of the parent compounds. RAC values are calculated as the relative concentrations of all chemicals in the last box of the river model and indicate the intrinsic potential of the different compounds for aquatic exposure. Because the same emission rate is used for all parent compounds, the RAC values are independent of the actual quantities used. For comparison within or across substance families, RACs can be normalized either to the RAC of each substance family's parent compound or to the RAC of atrazine as a well-known reference compound.

Figure 6 shows an example of model results for the substance family of mesotrione. In this case, the transformation products have a markedly higher mobility (manifested in lower values of the organic carbon-water partition coefficient K_{OC}) than the parent compound because mesotrione is cleaved into two smaller molecules (cyclohexane-1,3-dione (CHD) and 4-methylsulfonyl-2-nitrobenzoic acid (MNBA)). Whereas the relative masses in soil (m_S^{SS}) of the transformation products are lower than that of the parent compound, the higher mobility of the transformation products leads to higher fluxes (f_{SW}) into the water body compared to the parent compound. The RAC pattern, in contrast, is similar to that of the fluxes f_{SW}, because all compounds exhibit low reactivity in water. A similar gain in relative importance of the transformation products in the water compartment can be observed if the parent compound is neutral and transformation products are acidic compounds, or when a major structural change takes place such as the loss of a phosphate group that sorbs strongly in soils, which is, e.g., the case in glyphosate biodegradation.

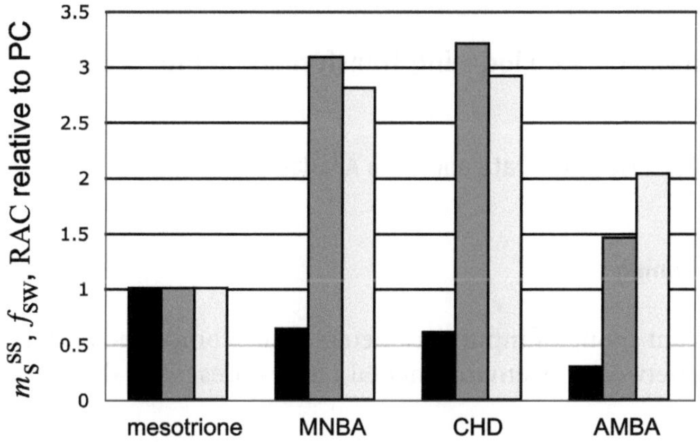

Fig. 6 Comparison of m_S^{ss} (steady-state mass in soil), f_{SW} (flux from soil to water) and RAC (relative aquatic concentration) within the mesotrione substance family (MNBA: 4-methylsulfonyl-2-nitrobenzoic acid, CHD: cyclohexane-1,3-dione, AMBA: 2-amino-4-methylsulfonylbenzoic acid). All values normalized to values of the parent compound. Reprinted with permission from [8], p 2448. © (2007) American Chemical Society

RAC values have been calculated for 16 pesticide and 53 of their experimentally identified transformation products [8]. Among the compounds with the 20 highest RAC values, there are 12 transformation products, which confirms the importance of transformation products as water pollutants. In this group of 20 compounds with highest RAC values, compounds from mainly five substance families are present: the chloroacetanilide alachlor, the triketones sulcotrione and mesotrione, and the acidic herbicides dicamba and bromoxynil-oct. The acidic transformation products of chloroacetanilide compounds (here alachlor OXA, ESA, and sulfinyl acetic acid) have a high potential for entry into surface waters, which is consistent with findings from monitoring studies [42, 44]. The transformation products of the triazine herbicide atrazine are not part of the top 20 list because they are not particularly mobile. However, because triazine herbicides are used in high amounts, their transformation products are nevertheless ubiquitous in surface and groundwater samples [45]. For sulcotrione, mesotrione, dicamba, and bromoxynil-oct, hardly any monitoring information exists. This discrepancy between the top 20 list and availability of monitoring data shows the potential of the RAC indicator to help identify relevant transformation products for future monitoring programs: Multiplication of RAC values with the amount of parent compounds used in a specific region of interest yields an indication of the expected relative concentrations of each transformation product and can thus inform chemical analysis about potential target compounds.

4 Data Requirements for Modelling Transformation Products

4.1 Compound-Specific Input Data and Data Availability

4.1.1 Phase Partitioning

One important group of input parameters is the compounds' partition coefficients, K_{xy}, between the environmental phases present in the model (Eq. 3).

$$K_{xy} = \frac{c_x}{c_y} \tag{3}$$

K_{xy} is the dimensionless partition coefficient of a compound between phases x and y, and c_x and c_y are the equilibrium concentrations of the compound in phases x and y, respectively.

The base set of partition coefficients needed for most multimedia models include the Henry's law constant (K_H) to describe partitioning between air and water, partition coefficients between water and various solid phases in soils, sediments, and particulate matter in the water column (K_d), and a coefficient describing partitioning between air-borne particles and air (K_p). If not available from direct experimental measurements, solid phase-water partition coefficients are often derived from the organic carbon-water partition coefficient, K_{oc} (Eq. 4), the underlying assumption being that sorption into organic matter dominates the overall sorption to bulk solid material.

$$K_d = f_{oc} \cdot K_{oc} = f_{oc} \cdot \frac{c_{oc}}{c_w} \tag{4}$$

with f_{oc} being the mass fraction of organic carbon in bulk solid material. If no experimental values for K_{oc} are available, it is usually estimated from the compound's octanol-water partition coefficient (K_{ow}) using a log-log-linear relationship [46]. Note, however, that the use of such relationships is questionable for polar compounds such as pesticides, pharmaceuticals and polar transformation products and the use of pp-LFERs has been suggested alternatively [47–51].

The availability of partition coefficients between and/or solubilities for water, organic matter and air, i.e., K_H and K_{oc}, S_w (water solubility), V_p (vapor pressure), and K_{ow}, is usually fairly good for most industrial compounds. However, partitioning data are scarce for transformation products. Whereas we were still able to find experimental K_{oc} values for 18 of the 53 pesticide transformation products introduced in Sect. 3.4, no such information is usually available for transformation products of other compound classes.

4.1.2
Degradation Half-Lives

The models further require information on degradation half-lives in each environmental compartment represented in the model. A minimum set usually comprises half-lives in soil, surface water, sediment, and air. Soil half-lives are mostly derived from dissipation studies in bulk soil and usually describe biodegradation and/or hydrolysis. Water half-lives are either obtained from simulation studies in water-sediment systems, derived from observed disappearance in natural water bodies, or calculated from separate information on hydrolysis, direct and indirect photolysis, and biodegradation. Half-lives in sediment are usually also obtained from simulation studies in water-sediment systems. However, it should be noted that these studies often do not generate enough data to clearly separate sediment and water half-lives. Half-lives in air mainly represent degradation due to reaction with OH radicals and can be measured in smog-chamber experiments, at least for relatively volatile chemicals. Note that when transformation product formation and dynamics are investigated, it is especially important that the half-lives describe primary degradation of the compound rather than complete mineralization.

Except for pesticides, some high-production-volume chemicals and, more recently, some pharmaceuticals and biocides, measured half-lives are usually scarce. Experimental half-lives for transformation products, except for soil half-lives of some well-known pesticide transformation products, are usually not available. Therefore most of the degradation information entered into multispecies multimedia models is estimated.

4.1.3
Transformation Schemes and Fractions of Formation

The third set of input data is specific to modelling transformation products. To do so, transformation schemes comprising the relevant transformation products, their connectivity, and fractions of formation for each transformation step are needed (for an example see Fig. 7). Depending on the purpose of the study, these transformation schemes may describe transformation up to mineralization in each compartment or they may describe formation of a few transformation products of specific interest.

For pesticides, transformation schemes in soil and water are often available in handbooks [52, 53] and in registration information from the EU [54], the U.S. [55], and the UK [56]. For all other compound classes, transformation schemes must be assembled from the available scientific literature. However, for most chemicals in commerce, information on possible transformation products is not available.

Fractions of formation as needed for the model algorithm described in Sect. 3.1 are usually not reported directly. They can, however, be calculated

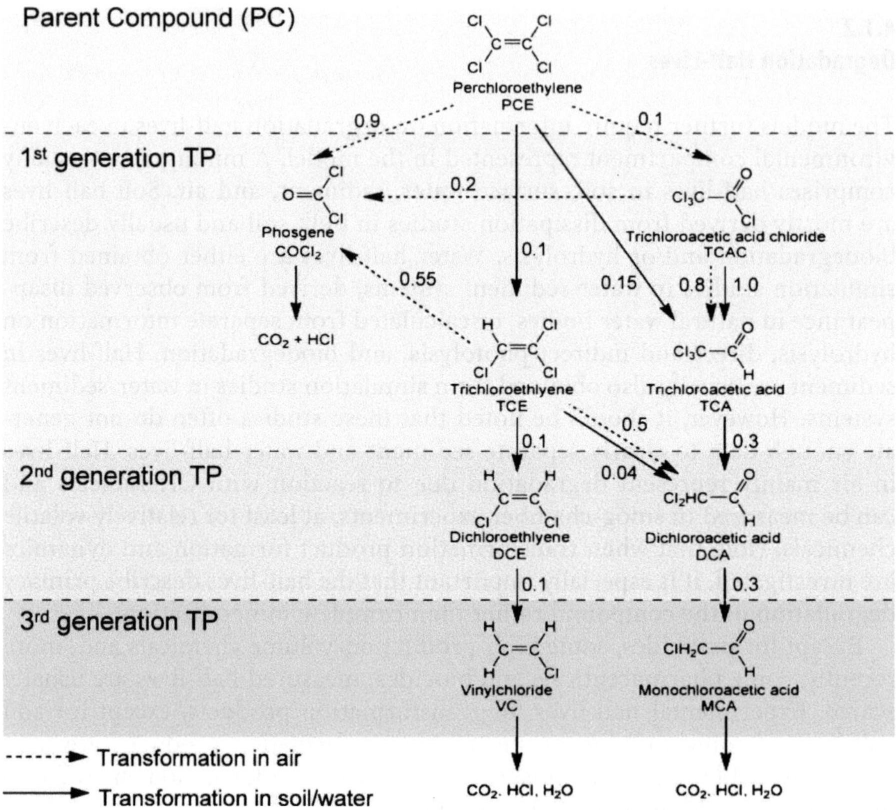

Fig. 7 Scheme of perchloroethylene degradation in soil, water, and air and corresponding fractions of formation. Reprinted with permission from [5], p 38. © (2003) Society for Risk Analysis

from information from degradation studies. Under the assumption of first-order kinetics, the time elapsed between the start of a degradation study and the time when the maximal concentration of transformation product j is reached ($t_{j,\max}$) can be expressed as a function of the degradation rate constants of precursor i (k_i) and transformation product j (k_j) (see Eq. 5). The maximum amount of a transformation product j formed ($c_{i,\max}$) is further a function of the fraction of formation (ff_{ij}) of transformation product j formed out of precursor i (see Eq. 6).

$$t_{j,\max} = \frac{\ln k_i - \ln k_j}{k_i - k_j} \tag{5}$$

$$c_{j,\max} = ff_{ij} \cdot c_{i,0} \left(\frac{k_i}{k_j} \right)^{\frac{k_j}{k_j - k_i}} . \tag{6}$$

Thus, if the half-life of the precursor in a given study is known, the fraction of formation and the half-life of the transformation product can be estimated from its maximal amount formed ($c_{j,\max}/c_{i,0}$) and the time it takes to reach this maximum ($t_{j,\max}$). While the degradation rate constant of the transformation product j (k_j) can then be deduced by numerically solving Eq. 5, the fraction of formation ff_{ij} can be calculated from Eq. 6.

However, such detailed information from degradation studies is very rare and fractions of formation therefore have to be estimated in most instances. Often, generic values of 1.0 for a single transformation product, 0.5 for two transformation products, and 0.33 for three transformation products have to be used; these values can be reduced by 10 or 20% to account for the fact that there are usually also some minor transformation products formed that are not explicitly accounted for in the modeled system.

4.2
Performance of Selected Property Estimation Software

4.2.1
Prediction of K_{OC} with KOCWIN and pp-LFERs

The two major approaches currently in use and under discussion [50, 57, 58] for the prediction of K_{OC} of neutral compounds are KOCWIN, which is part of the EpiSuite package [39], and different pp-LFER equations [48, 50, 58]. KOCWIN is a quantitative structure-activity relationship (QSARs) developed with molecular connectivity indices (MCI) [59]. pp-LFERs describe partitioning based on a few fundamental solute-bulk phase intermolecular interactions such as van-der-Waals interactions and H-bonding. The partitioning behavior of a given solute can thus be represented by a small set of descriptors (Abraham solvation parameters), which indicate its capacity for a set of defined intermolecular interactions.

In the context of transformation products, which are often oxidized and therefore more polar than their parent compounds, the performance of these approaches for polar compounds is of particular interest. This aspect has been discussed by Nguyen et al. [50] who compared the performance of KOCWIN with that of a newly deduced pp-LFER equation for a set of 75 compounds that span a log K_{OC} range from 1 to 7 and have experimentally determined Abraham parameters available for all of them. Nguyen et al. found mean errors of $\log K_{OC}$ of 0.18 for the pp-LFERs as compared to 0.2–0.43 for KOCWIN and different substance classes. The better performance of the pp-LFER was even more obvious when maximal prediction errors were considered: A maximal error of 0.48 log units for the pp-LFER predictions compared to a maximal error of 1.4 for the subclass of polar compounds as predicted by KOCWIN. Schüürmann et al. [57] compared the performance of KOCWIN with that of the pp-LFER from Poole and Poole [48] for a larger set of compounds ($n = 571$).

They confirmed that for polar compounds the pp-LFER approach was superior over KOCWIN in those cases where experimental Abraham parameters were available. If the Abraham parameters themselves, however, had to be predicted with a group contribution approach, the performance of the pp-LFER significantly deteriorated to the point where hardly any correlation between experimental and predicted values could be recognized. Own comparisons between KOCWIN predictions and experimental values for nine pesticides and eight pesticide transformation products yielded an average error of 0.93 log units translating into a factor of eight, which is considerably larger than the errors reported before. In conclusion, the following procedure for the estimation of K_{OC} values of neutral compounds seems currently most appropriate: If experimental Abraham parameters are available for a given compound, pp-LFERs should be used. In all other cases, the use of KOCWIN seems preferable; however, average errors of factors of three to eight and maximal errors as high as a factor of 30 must be expected.

For compounds with acidic or basic functional groups, speciation must additionally be accounted for when sorption coefficients are estimated. In this case, the apparent sorption coefficient, D_{OC}, is a composite of the sorption of at least two species (see Eq. 7).

$$D_{OC} = \alpha \cdot K_{OC,HA} + (1 - \alpha) \cdot K_{OC,A} \tag{7}$$

with HA and A being the protonated and deprotonated species, respectively, and α being the fraction of compound A in the protonated form at a given pH. Unfortunately, knowledge of how to predict sorption of cations and anions of organic compounds to organic matter or of how to relate it to the sorption behavior of the corresponding neutral species is scarce and, to our knowledge, no attempts to generalize respective experimental findings have been made so far. Based on our own comparisons of sorption data from the literature, we derived the following educated guesses to estimate sorption coefficients for cations and anions from that of the corresponding neutral species [8]: $\log K_{OC}(\text{anion}) = \log K_{OC}(\text{neutral}) - 1$ and $\log K_{OC}(\text{cation}) = \log K_{OC}(\text{neutral}) + 2$. We also suggest that for transformation products that are structurally strongly related to their parent compounds, and for whose parent compounds experimental K_{OC} values are available, an educated guess based on analogy reasoning might be preferable over the estimation procedures discussed.

4.2.2 Prediction of Half-Lives with BIOWIN

Currently, the model most often used for the prediction of biodegradation half-lives is the BIOWIN Primary Survey Model (PSM) from the EPISuite package [39]. This model is described in detail in the chapter by Howard. It is based on the results of an expert survey and uses a group contribution approach to predict biodegradability on a scale from 1 to 5. To convert this output into compartmental half-lives, it is suggested in the EPISuite package that the results from BIOWIN PSM be translated into water half-life categories (<1.75: 180 d, 1.75–2.25: 60 d, 2.25–2.75: 37.5 d, 2.75–3.25: 15 d, 3.25–3.75: 8.7 d, 3.75–4.25: 2.3 d, 4.25–4.75: 1.3 d, >4.75: 0.2 d) and that soil and sediment half-lives be derived from these water half-lives using conversion factors of two and nine, respectively. More recently, higher half-lives were suggested for BIOWIN scores below 2.25 [60]. Alternatively, Arnot et al. [61] suggest a regression equation for translating BIOWIN raw output into half-lives, which they have derived from experimental soil and water half-lives of a set of 40 diverse chemicals.

In the context of our work on the exposure assessment of pesticides and their transformation products [8], we evaluated how well their soil half-lives

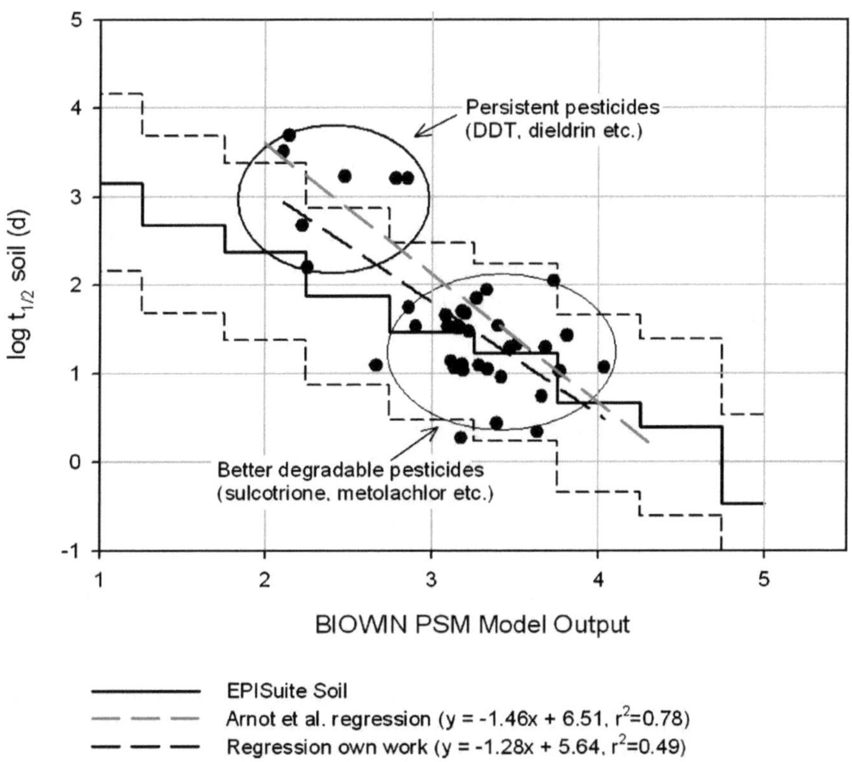

Fig. 8 Comparison of BIOWIN PSM output with experimental soil half-lives for 38 pesticides and pesticide transformation products. In addition, three possible methods for translating BIOWIN PSM output into actual half-lives are also indicated: EPISuite translation rules with modifications for PSM scores <2.25 as suggested in [60] (indicated as "EPISuite Soil"), the Arnot et al. [61] regression and our own regression based on the pesticide data. The finely *dashed lines* indicate uncertainty intervals of a factor of ten around the EPISuite translation rules. Reprinted with permission from [62], p 688. © (2006) Swiss Chemical Society

could be predicted with the BIOWIN Primary Survey Model (PSM). To translate the PSM output into half-lives, we used the translation rules suggested in the EPISuite package, the regression equation by Arnot et al. [61], as well as an own regression equation fitted to our collection of pesticide data (comprising 38 compounds with experimental soil half-lives, including 20 parent pesticides and 18 pesticide transformation products). Where more than one soil half-life was reported for a given compound, the median was used in the comparison. Figure 8 shows how the BIOWIN PSM output compares to the experimental soil half-lives, and how the EPISuite translation rules and the two regression equations compare to the data points.

It was found that the output of BIOWIN PSM and the median experimental half-lives do not correlate well ($r^2 = 0.49$). Also, for the better degradable pesticides with BIOWIN outputs between 2.5 and 4, little correlation between experimental soil half-lives and BIOWIN output could be detected. However, as can be seen in Fig. 8, with the exception of very recalcitrant pesticides such as DDT, heptachlor, aldrin and dieldrin, the predicted biodegradation half-lives in soil lie within a factor of ten of the measured half-lives. Further, the root mean squared errors between half-lives derived from the BIOWIN output and experimental half-lives indicate an average error of about a factor of five. From these comparisons of experimental and predicted half-lives, it can therefore be concluded that soil half-life predictions are usually accurate within a factor of five to ten. For a first assessment of the exposure potential of various transformation products, this accuracy might be considered sufficient.

4.2.3 Prediction of Transformation Schemes with CATABOL and UM-PPS

A critical factor currently limiting the exposure assessment of transformation products for substance classes other than pesticides is the lack of information on transformation schemes. Various computer tools exist whose goal it is to predict aerobic biodegradation pathways [63–66], but to the best of our knowledge, only CATABOL [65] (see chapter by Howard) and the University of Minnesota Pathway Prediction System (UM-PPS) (see chapter by Wackett & Ellis) [66] are presently under constant development. In both systems, pathway prediction is based on an extensive collection of known enzymatic reactions, based on which transformation rules have been compiled that translate a substrate substructure into a product substructure.

CATABOL and UM-PPS use different approaches to arrive at a prioritization of pathways. CATABOL uses information on the percentage of biological oxygen demand (%BOD) for the degradation of a given compound in a standard ready biodegradability test (OECD 301C). Using the transformation rules from UM-PPS and a large number of compounds with known test results, possible transformation pathways were generated and regressed against the known %BOD values. This procedure yielded probabilities for each transformation rule that can be used directly in the assignment of pathway priority. The mathematical algorithm of CATABOL further calculates the relative abundance of each predicted transformation product after a 28-day degradation period as well as their biodegradation half-lives. We tested CATABOL against the known transformation products of the 16 case study pesticides introduced in Sect. 3.2 [8] by running it over six generations and allowing for five parallel transformation pathways at each node. It was found that if the five transformation products with the highest relative mass were selected for each pesticide, they encompassed only 25% of known transformation products. If all of the 3900 predicted transformation products were examined, they were found to cover 60% of the 53 known transformation products. Thus, neither the accuracy nor the specificity of CATABOL is high enough to be useful for an efficient identification of possibly relevant transformation products.

UM-PPS, on the other hand, predicts all possible pathways that can be created from its transformation rules, and combines them with an indication of their degree of likelihood based on expert knowledge (see chapter by Wackett & Ellis). This approach is transparent and openly states that biodegradation pathway prediction will and should always be subject to some degree of expert judgment. However, UM-PPS is still in its infancy and suffers from an even more severe combinatorial explosion and hence lower specificity than CATABOL, i.e., the broad applicability of some of the rules leads to large numbers of transformation products and pathways after only a few generations. However, intensive work is currently ongoing to prioritize UM-PPS pathways based on thermodynamic considerations and structure-based rule refinement (see chapter by Wackett & Ellis), which is expected to yield considerably refined transformation rule priorities in the near future.

Both tools additionally suffer from the small number of reliable biodegradation studies in which transformation pathways were elucidated in enough detail to serve as training data for rule development and prioritization. Especially when confronted with compounds such as current-use pesticides and pharmaceuticals, both tools are lacking some of the rules necessary to break down these more complex molecular structures. Hopefully this situation will improve in the future since with the implementation of REACH, which requests identification of relevant transformation products for all compounds produced in amounts exceeding 100 t/year, the experimental biodegradation database is expected to grow considerably.

5 Conclusions and Outlook

As evidenced by monitoring studies and the modelling results presented in this chapter, transformation products do contribute to the overall burden of micropollutants in the environment. Transformation products are therefore not only an interesting object of scientific study but they also need to be acknowledged as a significant part of many pollution problems. However, the actual extent of exposure to transformation products remains unclear because the number of monitoring studies specifically addressing transformation products is currently still small and they are largely focused on the well-known pesticide transformation products. It can, however, be assumed that there are many more transformation products present in the environment that we are completely unaware of. Here, modelling can play an important role in pointing towards new, potentially relevant transformation products.

As discussed in this chapter, such modelling studies are, however, currently still subject to considerable uncertainty. As analyzed in detail in [5], the largest uncertainty in the prediction of Joint Persistence and exposure concentrations stems from uncertainty in the prediction of degradation half-lives, which amounts to at least a factor of 5–10. Partition coefficients generally exhibit somewhat lower uncertainties and also tend to be less influential. Fractions of formation, although highly uncertain, are bounded between 0 and 1. As long as the focus is on major transformation products (>10%), the uncertainty in fractions of formation is therefore always below a factor of ten.

An even more fundamental uncertainty is introduced by the prediction of transformation products in those cases where no experimental evidence about possible degradation pathways is available. As pointed out earlier, such pathway prediction tools are still in their infancy and not readily applicable to predict a manageable set of likely transformation products.

To further advance knowledge on the presence and relevance of transformation products, a close interplay of modelling and experimental studies is advocated. On the monitoring side, more exploratory studies with the aim of identifying new transformation products are needed and have recently become more feasible with the advent of high-resolution mass spectrometry. Also, field studies should not just focus on the presence or absence of transformation products but also elucidate their transport and fate mechanisms in the environment. Finally, more systematic experimental investigations of degradation reactions are urgently needed, particularly in order to expand the training sets for degradation estimation methods. The results of these field and laboratory studies should be closely linked to further model development. With the investigation of further case studies, uncertainties in the models will become more transparent and the models more amenable to further refinement.

Once the procedure for investigating and modelling of transformation products has become clearer, parts of it may be generalized so that application to more diverse cases becomes possible. This should also ultimately make it possible to put more stringent requirements on the inclusion of transformation products in chemical risk assessment in the context of product authorization and regulation.

Acknowledgements We thank Konrad Hungerbühler and René P. Schwarzenbach for their continuing support of our research on transformation products, and Heinz Singer and Juliane Hollender for the many fruitful discussions on the topic. The Swiss Federal Office for the Environment (FOEN) funded parts of this research through the projects Mikropoll I and KoMet.

References

1. Bouchard M, Gosselin NH, Brunet RC, Samuel O, Dumoulin MJ, Carrier G (2003) Toxicol Sci 73:182
2. Belfroid AC, van Drunen M, Beek MA, Schrap SM, van Gestel CAM, van Hattum B (1998) Sci Total Environ 222:167
3. Sinclair CJ, Boxall ABA, Parsons SA, Thomas MR (2006) Environ Sci Technol 40:7283
4. Fenner K, Scheringer M, Hungerbühler K (2000) Environ Sci Technol 34:3809
5. Fenner K, Scheringer M, Hungerbühler K (2003) Risk Anal 23:35
6. Fenner K (2001) PhD Thesis, ETH Zurich, Zurich, Switzerland
7. Fenner K, Kooijman C, Scheringer M, Hungerbühler K (2002) Environ Sci Technol 36:1147
8. Gasser L, Fenner K, Scheringer M (2007) Environ Sci Technol 41:2445
9. Quartier R, Müller-Herold U (2000) Ecol Model 135:187
10. Cahill TM, Cousins I, Mackay D (2003) Environ Toxicol Chem 22:483
11. Cahill TM, Mackay D (2003) Chemosphere 53:571
12. Gandhi N, Bhavsar SP, Gewurtz SB, Diamond ML, Evenset A, Christensen GN, Gregor D (2006) Environ Sci Technol 40:4714
13. Schenker U, Scheringer M, Hungerbühler K (2007) Environ Sci Poll Res 14:145
14. FOCUS (2006) http://viso.jrc.it/focus/
15. Ramanarayanan T, Narasimhan B, Srinivasan R (2005) J Agric Food Chem 53:8848
16. Lindstrom A, Buerge IJ, Poiger T, Bergqvist PA, Muller MD, Buser HR (2002) Environ Sci Technol 36:2322
17. Jonkers N, Laane R, de Graaf C, de Voogt P (2005) Estuar Coast Shelf Sci 62:141
18. Ma QL, Wauchope RD, Ma L, Rojas KW, Malone RW, Ahuja LR (2004) Pest Manage Sci 60:267
19. Mackay D (ed) (2001) Multimedia Environmental Models. The Fugacity Approach. Lewis Publishers, Boca Raton, FL
20. Scheringer M (2002) Persistence and Spatial Range of Environmental Chemicals. Wiley, Weinheim
21. Muir DCG, Howard PH (2006) Environ Sci Technol 40:7157
22. Klasmeier J, Matthies M, Macleod M, Fenner K, Scheringer M, Stroebe M, Le Gall AC, McKone T, Van De Meent D, Wania F (2006) Environ Sci Technol 40:53
23. OECD (2002) OECD guideline for the testing of chemicals (OECD 307). OECD. Paris
24. OECD (2002) OECD guideline for the testing of chemicals (OECD 308). OECD. Paris
25. UNEP (2001) In: Final Act of the Conference of Plenipotentiaries on The Stockholm Convention On Persistent Organic Pollutants. United Nations Environment Program. Geneva, Switzerland
26. Scheringer M (1996) Environ Sci Technol 30:1652
27. Scheringer M (1997) Environ Sci Technol 31:2891
28. Scheringer M, Wania F (2003) In: Fiedler H (ed) The Handbook of Environmental Chemistry, vol 3. Springer, Berlin Heidelberg

29. Wania F, Mackay D (1995) Sci Total Environ 161:211
30. Scheringer M, Wegmann F, Fenner K, Hungerbühler K (2000) Environ Sci Technol 34:1842
31. Wegmann F, Scheringer M, Moller M, Hungerbühler K (2004) Environ Sci Technol 38:1505
32. Semeena S, Lammel G (2003) Fresen Environ Bull 12:925
33. Bennett DH, McKone TE, Matthies M, Kastenberg WE (1998) Environ Sci Technol 32:4023
34. MacLeod M, Mackay D (2004) Environ Poll 128:241
35. Wania F (2003) Environ Sci Technol 37:1344
36. Wania F (2004) GAIA 13:176
37. Howard PH (1991) Handbook of Environmental Fate and Exposure Data for Organic Chemicals, vol III. Pesticides. Lewis, Chelsea, Michigan
38. Mackay D, Shiu WY, Ma KC (1997) Illustrated Handbook of Physical-Chemical Properties and Environmental Fate for Organic Chemicals. Lewis Publishers, Boca-Raton, FL
39. EPA US (2004) In: http://www.epa.gov/opptintr/exposure/pubs/episuitedl.htm (ed). Syracuse Research Corporation
40. Hung H, Blanchard P, Halsall CJ, Bidleman TF, Stern GA, Fellin P, Muir DCG, Barrie LA, Jantunen LM, Helm PA, Ma J, Konoplev A (2005) Sci Total Environ 342:119
41. Krueger HU, Gavrilov R, Liu Q, Zetzsch C (2005) Development of a method for determining tropospheric degradation of semivolatile pesticides by OH radicals. Deutsches Umweltbundesamt, Berlin
42. Rebich RA, Coupe RH, Thurman EM (2004) Sci Total Environ 321:189
43. Kalkhoff SJ, Lee KE, Porter SD, Terrio PJ, Thurman EM (2003) J Environ Qual 32:1025
44. Battaglin WA, Thurman EM, Kalkhoff SJ, Porter SD (2003) J Am Water Resour Assoc 39:743
45. Gilliom RJ, Barbash JE, Crawford CG, Hamilton PA, Martin JD, Nakagaki N, Nowell LH, Scott JC, Stackelberg PE, Thelin GP, Wolock DM (2006) The Quality of Our Nation's Waters – Pesticides in the Nation's Streams and Ground Water, 1992–2001. U.S. Geologial Survey (eds), Reston, Virginia
46. Karickhoff SW (1984) J Hydraulic Eng 110:707
47. Abraham MH, Andonianhaftvan J, Whiting GS, Leo A, Taft RS (1994) J Chem Soc-Perkin Trans 2:1777
48. Poole SK, Poole CF (1999) J Chromatogr A 845:381
49. Goss KU, Schwarzenbach RP (2001) Environ Sci Technol 35:1
50. Nguyen TH, Goss KU, Ball WP (2005) Environ Sci Technol 39:913
51. Götz CW, Scheringer M, MacLeod M, Roth CM, Hungerbühler K (2007) Environ Sci Technol 41:1272
52. Roberts TR (ed) (1998) Metabolic pathways of agrochemicals, vol Part 1. The Royal Society of Chemistry, Cambridge
53. Roberts TR (ed) (1999) Metabolic pathways of agrochemicals, vol Part 2. The Royal Society of Chemistry, Cambridge
54. European Commission (2006) Plant protection products. http://europa.eu.int/comm/food/plant/protection/evaluation/new_subs_rep_en.htm
55. U.S. Environmental Protection Agency (2006). Pesticides: Reregistration. http://cfpub.epa.gov/oppref/rereg/status.cfm?show=rereg
56. U.K. Pesticides Safety Directorate (2006). http://www.pesticides.gov.uk/psd_evaluation_all.asp
57. Schüürmann G, Ebert RU, Kuehne R (2006) Chimia 60:691

58. Niederer C, Goss KU, Schwarzenbach RP (2006) Environ Sci Technol 40:5374
59. Meylan W, Howard PH, Boethling RS (1992) Environ Sci Technol 26:1560
60. Aronson D, Boethling RS, Howard PH, Stiteler W (2006) Chemosphere 63:1953
61. Arnot J, Gouin T, Mackay D (2005) Practical methods for estimating environmental biodegradation rates. Canadian Environmental Modelling Network
62. Fenner K, Canonica S, Escher BI, Gasser L, Spycher S, Tulp HC (2006) Chimia 60:683
63. Punch B, Patton A, Wight K, Larson B, Masscheleyn P, Forney L (1996) In: Peijnenburg WJGM, Damborsky J (eds) Biodegradability Prediction. Kluwer, Dordrecht
64. Klopman G, Tu MH (1997) Environ Toxicol Chem 16:1829
65. Jaworska J, Dimitrov S, Nikolova N, Mekenyan O (2002) SAR QSAR Environ Res 13:307
66. Hou BK, Ellis LBM, Wackett LP (2004) J Ind Microbiol Biotechnol 31:261

Hdb Env Chem Vol. 2, Part P (2009): 151–174
DOI 10.1007/698_2_017

Published online: 9 April 2009

Treatment of Transformation Products

Craig D. Adams

Dept. of Civil, Environmental, and Architectural Engineering, University of Kansas, Lawrence, KS 66045, USA
cadams@ku.edu

Abstract Drinking water and wastewater treatment processes play an important role regarding formation and removal of transformation products. As such, treatment processes directly impact human and environmental health risks, as they affect both exposure and toxicity of the pool of synthetic organic compounds. In this chapter, the key drinking water and wastewater processes that may cause further transformation of transformation products are reviewed and prioritized, as are the key partitioning mechanisms.

Keywords Disinfection · Partitioning · Sorption · Transformation · Wastewater treatment · Water treatment

Abbreviations

ATZ	Atrazine
DDA	Didealkyatrazine
DEA	Deethylatrazine
DI	Distilled water
DIA	Deisopropylatrazine
ETBE	Ethyl-*tert*-butyl ether
GAC	Granular activated carbon
k'	Pseudo-first-order rate constant
K_C	Half-maximum growth rate
K_D	Linear isotherm coefficient
K_{OW}	Octanol/water partition coefficient
MCA	Monochloramine
MLSS	Mixed liquor suspended solids (or biomass)
MTBE	Methyl-*tert*-butyl ether
ORP	Oxidation–reduction potential
PAC	Powdered activated carbon
PACl	Polyaluminum chlorides
TBA	*Tert*-butyl alcohol
TBF	*Tert*-butyl formate
TP	Transformation product
U_{max}	Microorganism's maximum growth rate
UV	Ultraviolet
Y	Yield coefficient

1 Introduction

In preceding chapters, the fate and occurrence of transformation products of synthetic organic chemicals in the environment have been examined, including mechanisms, models, and validation. Both wastewater and drinking water treatment have significant relevance with regard to the transformation products of pesticides, solvents, gasoline, fuel oxygenates, surfactants, pharmaceuticals, personal care products, and other synthetic chemicals.

During wastewater treatment, transformation products may be formed through both abiotic and biotic reactions, and removed through additional transformations as well as partitioning mechanisms. Many of these degradates may, instead, be discharged in the sewage outfall, while others might reside in the residual sludge. Thus, wastewater treatment might serve as a source or input of transformation products into the environment. Alternate inputs of transformation products into the environment include spills, other point-source discharges, back-siphoning, non-point source applications, and other sources.

During drinking water treatment, the source of transformation products may be either from the influent source water to the treatment plant (i.e., a lake, river, or groundwater), or, alternatively, from abiotic or biotic reactions in the treatment process itself. Transformation products may also undergo further transformation in a drinking water treatment plant. Alternatively, these products may, instead, partition onto treatment solids or treatment chemicals (e.g., activated carbon), thereby, being removed from the treated water. Drinking water treatment has the potential to either decrease or increase concentrations of specific transformation products, which are initially present in the source water, through combinations of chemical reactions and partitioning reactions.

For reasons discussed in preceding chapters, there is a concern, as well as an uncertainty, regarding the identity and concentrations of the wide array of transformation products that are entering the environment (including drinking water supplies) from wastewater treatment plants and other sources. There are even more uncertainties with respect to which transformation products are commonly formed within water treatment plants, as well as their fate after formation. Detailed study of the formation, identity, and fate of pesticide transformation products in drinking water treatment is of considerable interest and is the subject of ongoing research (e.g., Adams et al. 2007).

The environmental risk and human-health risk of these transformation products are related, in both cases, to a combination of exposure and toxicity. Each synthetic organic chemical entering a treatment plant could form many different transformation products via oxidation, hydrolysis, reduction, and/or biodegradations. Thus, thousands of transformation products are possible, many of which are difficult or impossible to analyze, and most of which have not yet been identified analytically. Sinclair et al. (2006) addressed this issue for pesticides used in the United Kingdom (UK) and the United States (US) by prioritizing the risk associated with pesticide transformation products. The pesticide degradates that have been identified with the highest risk are tabulated in Tables 1 and 2 for the UK and US, respectively.

Table 1 Top 24 degradates with respect to risk based on usage, toxicity and transformation product formation, mobility, and persistence for the UK (Sinclair et al. 2006)

Parent	$\log K_{OC}$ (pH 7, 25 °C)	Degradate	$\log K_{OC}$ (pH 7, 25 °C)	ADI ($mg\,kg^{-1}\,d^{-1}$)	Risk index
Isoproturon	2.6	1-Methyl-3-(4-isopropyl phenyl)-urea	ND	0.015	3.458
Didofop-methyl	3.7	Diclofob acid	0.2	0.001	2.653
Tri-allate	4.7	TCPSA	ND	0.005	1.916
Propachlor	2.6	Propachlor oxanilic acid	ND	0.009	1.539

Table 1 (continued)

Parent	log K_{OC} (pH 7, 25 °C)	Degradate	log K_{OC} (pH 7, 25 °C)	ADI ($mg\,kg^{-1}\,d^{-1}$)	Risk index
Chlorothalonil	2.9	3-Carbamyl-2,4,5-trichloro-benzoic acid	ND	0.018	0.98
Propachlor	2.6	Propachlor ethane sulfonic acid	ND	0.009	0.883
Methiocarb	2.9	Methiocarb sulfoxide	1.7	0.002	0.851
Chlorothalonil	2.9	4-Hydroxy-2,5,6-trichloro-isophthalonitrile	0.2	0.018	0.722
Chlorpyrifos/ triclopyr	4.0	3,5,6-Trichloro-2-pyridinol	0.0	0.003/0.005	0.69
Isoproturon	2.6	Desmethylisoproturon	2.7	0.015	0.615
Triazamate	2.4	Triazamate metabolite II	ND	0.0003	0.367
Atrazine	2.8	Deethylatrazine	2.2	0.006	0.277
Simazine	2.6	Deisopropylatrazine	2.0	0.005/0.006	0.201
Thifensulfuron-methyl	0.2	Thiophene sulfonimide	ND	0.01	0.106
Trifloxystrobin	4.2	CGA-321113	0.1	0.038	0.091
Thifensulfuron-methyl	0.2	Thifensulfuron acid	0.0	0.01	0.066
thiophanate-methyl	1.5	Carbendazim	2.1	0.006–0.03	0.066
Benomyl fluquin-conazole	ND				
Tebuconazole	3.3				
Tetraconazole	3.1	1,2,4-Triazole	ND	0.004–0.1	0.044
Propiconazole	3.5				
Myclobutanil	2.9				
Kresoxim-methyl	3.7	Kresoxim-methyl acid	ND	0.4	0.019
Desmedipham	3.3	Ethyl-*m*-hydroxyphenyl carbamate	2.4	0.0018	0.008
Tribenuron-methyl	0.2	Triazine amine A	2.0	0.12	0.004
Amitraz	4.4	BTS 27919	0.5	0.0025	0.004
Thifensulfuron-methyl	0.2	*O*-desmethyl thifen-sulfuron-methyl	ND	0.01	0.002
Picolinafen	3.5	CL 153815	0.0	0.014	0.001

ND = No data;
ADI = Allowable daily intake;
log K_{OC} calculated using Advanced Chemistry Development (ACD/Labs) Software V8.14 for Solaris (1994–2008 ACD/Labs)

In this chapter, the key processes that affect the formation and fate of transformation products in wastewater treatment and drinking water treatment are discussed.

Table 2 Top 16 degradates with respect to risk based on usage, toxicity and transformation product formation, mobility, and persistence for the USA (Sinclair et al. 2006)

Parent	K_{OC} (pH 7, 25 °C)	Degradate	K_{OC} (pH 7, 25 °C)	ADI ($mg\,kg^{-1}\,d^{-1}$)	Risk index
Diazinon	3.4	Pyrimidinol	1.8	0.002	6.695
Chlorpyrifos	4.0	3,5,6-Trichloro-2-pyridinol	0.0	0.003	3.549
Metam-sodium	0.0	Methylisothiocyanate	1.9	0.01	2.671
Diuron	2.9	*N′*-(3,4-dichlorophenyl)-*N*-methylurea	ND	0.007	1.635
Chlorpyrifos	4.0	3,5,6-Trichloro-2-methoxypyridine	3.4	0.003	1.446
Eethephon	0.0	2-Hydroxyethyl phosphonic acid	ND	0.018	0.711
Simazine	2.6	Deisopropylatrazine	2.0	0.005	0.557
Carbaryl	2.7	1-Napthol	ND	0.003	0.41
Malathion	3.0	Malathion dicarboxylic acid	0.0	0.05	0.289
Thiophanate-methyl	1.5	Carbendazim	2.1	0.02	0.08
Captan	2.4	Tetrahydrophthalamide	ND	0.1	0.043
Glyphosate	0.0	Aminomethylphosphonic acid	0.0	0.3	0.007
Aldicarb	2.0	Aldicarb sulfoxide	0.8	0.003	0.001
Iprodione	3.0	RP 30228	2.7	0.02	0.001
Aldicarb	2.0	Aldicarb sulfone	1.1	0.003	0.001
Cypermethrin	4.8	3-Phenoxybenzoic acid	0.7	0.015	0.001

ND = No data;
ADI = Allowable daily intake;
$\log K_{OC}$ calculated using Advanced Chemistry Development (ACD/Labs) Software V8.14 for Solaris (1994–2008 ACD/Labs)

2 Fate of Transformation Products in Drinking Water Treatment

2.1 Drinking Water Treatment Processes and Operation

Several typical water treatment plant schemes are shown in Figs. 1, 2, and 3 that are used to treat: (1) turbid surface water using coagulation/flocculation/sedimentation (Fig. 1), (2) groundwater using lime/soda ash softening (Fig. 2), and (3) groundwater using membranes (Fig. 3), respectively. In such systems, a wide suite of transformation products (and their synthetic organic chemical parents) may be in the surface water or groundwater being treated. In conventional surface water treatment, the key treatments include the addition of a coagulant, to destabilize the colloidal suspension, followed by gentle mixing (flocculation) and settling of the solids (Fig. 1). The

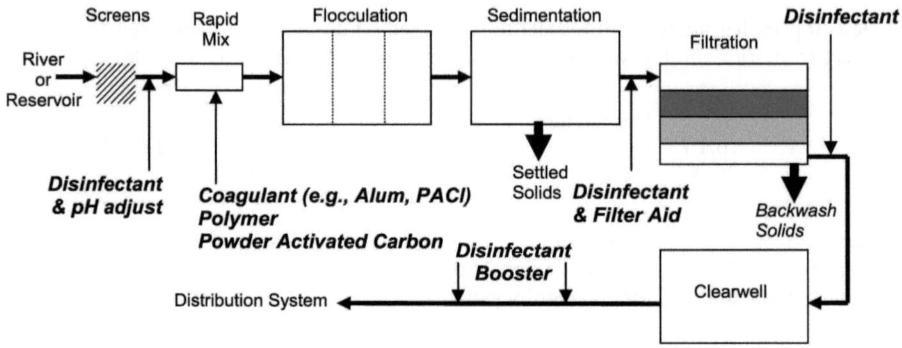

Fig. 1 A typical water treatment plant for treating turbid surface water

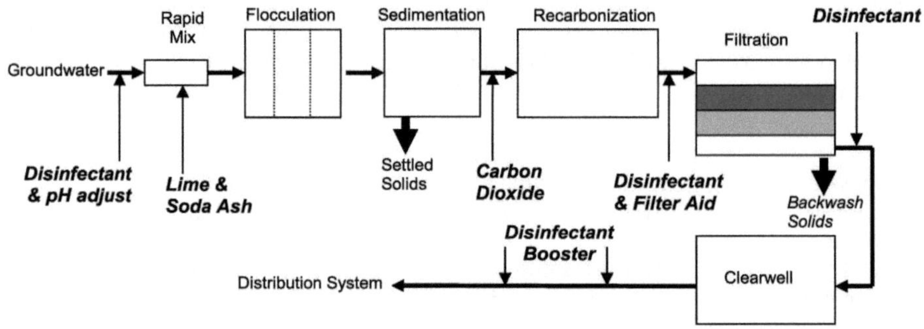

Fig. 2 A typical water treatment plant for treating groundwater, using lime softening

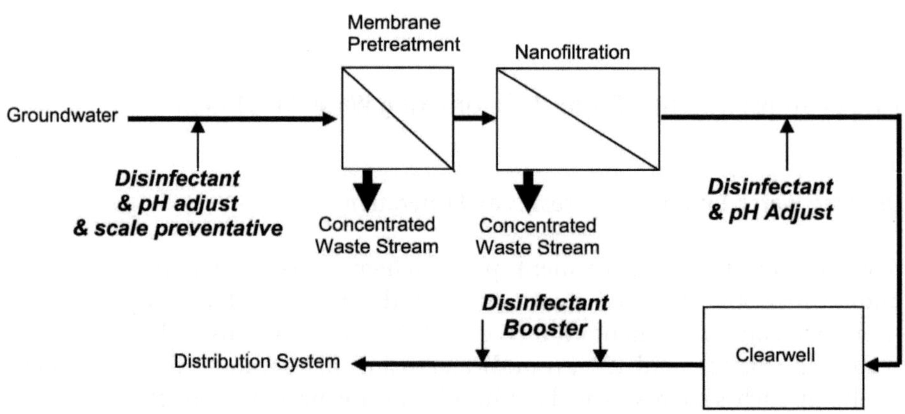

Fig. 3 A typical water treatment plant for treating groundwater, using membranes

remaining colloidal solids are removed via granular filtration. A variety of oxidants may be added during treatment at the inlet, prior to filtration, after filtration, and throughout the distribution system. These may include free chlorine ($HOCl/OCl^-$), monochloramine (MCA), chlorine dioxide (ClO_2), permanganate (MnO_4^-), and ozone (O_3), causing further transformation of the transformation products.

In a softening plant, hardness ions (e.g., Ca^{2+} and Mg^{2+}) are removed by precipitation of $CaCO_3$ and $Mg(OH)_2$ at pH levels of nominally 10.3 and 11.0, respectively (Fig. 2). Lime is added to raise the pH to the necessary levels, and soda ash (Na_2CO_3) is added if insufficient carbonate is present to precipitate the calcium hardness. Filtration and oxidants are also used as described above.

Membrane treatment plants have recently become more common, with a wide variety of membrane types and configurations being used (Fig. 3). Each membrane system will generally utilize pretreatment to remove large particles, to adjust pH, disinfect, and/or minimize scaling.

2.2
Abiotic Transformations in Drinking Water Treatment

There are opportunities for transformation products (and parent synthetic organic compounds) from the source waters to react abiotically within a water treatment plant. The most common abiotic reactions include chemical oxidation, hydrolysis, and photolysis.

2.2.1
Chemical Oxidation in Drinking Water Treatment

Chemical oxidants that are commonly used to disinfect, control taste and odor, remove color, and/or provide microflocculation, include free chlorine ($HOCl/OCl^-$), monochloramine (MCA), chlorine dioxide (ClO_2), permanganate (MnO_4^-), and ozone (O_3). Although the relative reactivity of these compounds varies greatly, very few studies have comprehensively examined their reactivity with transformation products. One comprehensive study showed that the relative reactivity of pesticides with these oxidants was, on average (Adams et al. 2007)

$$O_3 > HOCl/OCl^- > ClO_2 \sim MCA \sim MnO_4^- .$$

Specifically, for 39 different pesticides, 32 and 18% were highly reactive (> 50% removal of typical disinfectant exposures) for ozone and free chlorine, respectively (Adams et al. 2007). For monochloramine, chlorine dioxide, and permanganate, only 6% or fewer were highly reactive (Adams et al. 2007). While these reactivities were those of the parent pesticides (and not their transformation products), it would be anticipated that the relative reactivity of transformation products towards these oxidants would be similar, on average, i.e.,

$$O_3 > HOCl/OCl^- > ClO_2 \sim MCA \sim MnO_4{}^- .$$

With respect to transformation products existing in the environment, and, hence, in the influent to treatment plants, more study is needed to assess their reactivity towards this suite of drinking water oxidants.

It is instructive to consider relative reaction rates of parents and first- and second-oxidation byproducts as they relate to the buildup of transformation products. If the first transformation products of ozone (or any other oxidant) are more reactive than the parent compound, i.e.,

$$\text{Parent} + O_3 \xrightarrow{\text{slow}} \text{1st Products} + O_3 \xrightarrow{\text{fast}} \text{Other Products} ,$$

then the concentration of the transformation product(s) will not tend to increase. However, if the transformation products of ozone are less reactive than with the parent, i.e.,

$$\text{Parent} + O_3 \xrightarrow{\text{fast}} \text{1st Products} + O_3 \xrightarrow{\text{slow}} \text{Other Products} ,$$

then the concentration of the transformation product will tend to increase and may pose a potential health issue.

Ozone

An instructive example is the reaction of ozone with triazines. Ozone readily reacts with atrazine (2-chloro-4-ethylamino-6-isopropylamino-*s*-triazine), the most commonly used herbicide in the United States, to form deethylatrazine (DEA, 2-amino-4-chloro-6-isopropylamino-*s*-triazine) and deisopropylatrazine (DIA, 2-amino-4-chloro-6-ethyamino-*s*-triazine) in water treatment plants (Adams and Randtke 1992). These same two degradates are also formed through biologically mediated reactions in the soil (Adams and Thurman 1991), and are commonly found in groundwater and surface water (Jiang et al. 2006). Whether formed in the environment, or in a water treatment plant, ozone can further degrade the deethyl- and deisopropyl-atrazine transformation products to didealkylatrazine (DDA, 2-chloro-4,6-dialkyl-*s*-triazine) (Adams and Randtke 1992; Jiang and Adams 2006).

Ozone has a rate constant with atrazine of just $6.0\,\text{l}\,\text{mol}^{-1}\,\text{s}^{-1}$ (Acero et al. 2000)

$$\frac{dA}{dt} = -\,k_{O_3,A}\,[O_3]\,[A] .$$

The ozone rate constants for deethylatrazine and deisopropylatrazine are even slower, 0.18 and $3.1\,\text{l}\,\text{mol}^{-1}\,\text{s}^{-1}$, respectively, while the ozone rate constant for didealkylatrazine is slower yet at $< 0.1\,\text{l}\,\text{mol}^{-1}\,\text{s}^{-1}$ (Acero et al. 2000). Thus, ozonation of the transformation products would lead to further transformation, albeit with the potential for building up in concentration. A study by Jiang and Adams (2006) showed that while ozone is partially reactive with atrazine, the buildup of degradation products, the total chloro-*s*-triazine (TCT) (or sum of each constituent) decreases much less than the parent due to transformation product buildup (Fig. 4). This finding is consistent with Acero et al. (2000). Further, this effect was seen in a pilot-scale study at a full-scale surface water treatment plant where 4 mg/l of ozone reduced 5.6 μg/l of atrazine to 0.3 μg/l, and formed much higher concentrations of deethylatrazine (1.2 μg/l) (Hulsey et al. 1993).

Mascolo et al. (2001a) studied the ozonation of isoproturon, a phenylurea derivative. The study showed the parent compounds could be completely degraded by molecular ozone under typical drinking water treatment conditions. Many degradation products were formed and identified including complex structures maintaining the aromatic ring, as well as simpler transformation products, such as simple organic acids and aldehydes. The transformation products were further degradable by molecular ozone and/or hydroxyl radicals formed during ozonation.

Another example of transformation products being formed, and then further transformed, is the formation of *tert*-butyl alcohol (TBA) and *t*-butyl formate (TBF) from ethyl-*tert*-butyl ether (ETBE) or methyl-*tert*-butyl ether (MTBE) in hydroxyl-radical-mediated reactions during ozonation (Acero et al. 2001; Sutherland et al. 2005). These transformation products are subsequently oxidized to simpler organic compounds such as formic acid and acetic acid. Removal of these two transformation products (TBA and TBF)

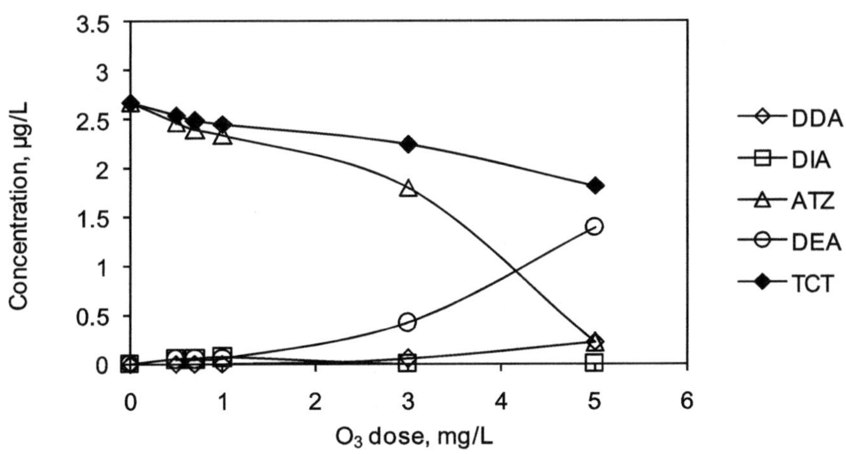

Fig. 4 Removal of atrazine (ATZ) and formation of deethylatrazine (DEA), deisopropylatrazine (DIA), and didealkylatrazine (DDA) using ozonation of 3 μg/l of atrazine in filtered Missouri River water at pH 6 (Jiang and Adams 2006)

requires about twice the ozone dosage as is required to remove the parent compound, while mineralization of the MTBE (to CO_2) requires much greater oxidant dosages.

Dantas et al. (2007) studied the formation of transformation products during ozonation of bezafibrate. The study showed that the transformation products were themselves further transformed by ozone.

Chlorine

Free chlorine is reactive with some pesticide parents, but not others. While a number of studies have examined formation of chlorination byproducts during disinfection, few studies have addressed subsequent removal of the degradates (e.g., Lopez et al. 1998). For example, Jiang and Adams (2006) determined that free chlorine at typical disinfection conditions in water treatment was ineffective at removal of deethylatrazine, deisopropylatrazine, and didealkylatrazine (as well as the parent compounds).

Mascolo et al. (2001b) found that chlorination of isoproturon at typical conditions caused the formation of a variety of transformation products of unknown toxicity. These transformation products were themselves partially degraded via subsequent chlorination.

Other work by Buth et al. (2007) examined formation, and subsequent removal of transformation products of cimetidine (an antacid). These degradates were shown to be more toxic overall than the parent compound, making them particularly problematic with respect to human health risk. The transformation products were also shown to be further degradable via chlorination to a degree highly dependent on chlorine dosage and pH.

2.2.2 Hydrolysis in Drinking Water Treatment

Another process causing degradation of transformation products (or parent compounds) is acid, neutral, and alkaline (or base) hydrolysis. While natural waters tend to have pH levels of between 6 and 9, much higher pH levels (approximately 10.3 and 11.0) are common in treatment plants during straight lime (calcium only) and excess lime (calcium plus magnesium) softening, respectively.

While hydrolysis products may be formed in the environment at lower pH levels, once in the treatment plant, there is significant potential for additional hydroxide-catalyzed hydrolysis transformation products to occur at higher pH levels. Few studies have examined the hydrolysis of transformation products in detail, and additional research is needed. Recent screening studies (e.g., Adams et al. 2007) have shown that many pesticides are readily hydrolyzable at the elevated pH levels that occur during softening operations in water treatment.

2.2.3 Photolysis in Drinking Water Treatment

Another potential abiotic reaction for transformation products is photolysis during ultraviolet (UV) disinfection. However, significant direct photolysis of low- and sub-μg/l concentrations of transformation products would generally be unlikely in water treatment, for several reasons. First, relatively low UV doses are typically used for disinfection in water treatment plants (e.g., 30–40 mJ/cm^2) (Crittenden et al. 2005), while higher dosages are required for wastewater treatment.

At these dosages for drinking water, even parent compounds would not be likely to be photolyzed to any great extent. For example, all but one of 39 pesticides had very low reactivity (and 0% of the pesticides were highly reactive), even with UV exposures well above typical levels in clean lab water, as demonstrated by Adams et al. (2007). Similarly, earlier work by Adams et al. (2002) concluded that negligible photolysis of antibiotics was observed in either lab water or natural waters.

A study by Sharpless et al. (2003) was able to achieve 95% removal of the parent compound, atrazine, during UV disinfection with a medium-pressure lamp, but only with approximately 100 times a typical UV dosage. A similar lack of removal of the most common transformation products of atrazine would be expected due to similar UV absorption characteristics.

Photolysis of transformation products in natural water containing natural organic matter would generally be expected to be negligible due to competitive absorbance effects. Specifically, at very low sub-μg/l concentrations of transformation products, only a very small percentage of the UV energy could be absorbed in the presence of much higher (i.e., mg/l) concentrations of strongly absorbing humic components of the natural organic matter.

The transformation products entering a treatment plant from surface water have often been exposed to sunlight in the environment. Thus, transformation products that are photochemically stable have, to some degree, been selected for in the natural environment. While few studies have examined the issue, direct photochemical conversion of transformation products in drinking water treatment would usually be minimal.

Combined UV Processes

It should be noted that use of UV in combination with hydrogen peroxide, ozone, or titanium dioxide creates highly reactive hydroxyl radicals. These advanced oxidation processes are highly effective at removal of many, or most, synthetic organic compounds, including transformation products of the parent. For example, Doll and Frimmel (2004) studied the formation and removal of transformation products of iopromide, iomeprol, clofibric acid, and carbamazepine by a simulated and actual solar/TiO_2

process. For clofibric acid, transformation products (including isobutyric acid, 2-(4-hydroxyphenoxy)-isobutyric acid, and hydroquinone) were subsequently degradable by the process to simpler organic structures and, eventually, CO_2.

Vogna et al. (2004) showed that UV/H_2O_2 (as well as molecular ozone) was effective at oxidizing transformation products of the pharmaceutical, diclofenac. Lau et al. (2007) studied the formation of a suite of ten transformation products of butylated hydroxyanisole by including 1,4-benzoquinone, *t*-butyl-1,4-benzoquinone, and hydroquinone. Some of the degradates were precipitated from solutions as an orange-colored solid that could be removed by filtration. The study showed that UV/ozone and ozonation were more effective than UV alone at removal of the parent and transformation products.

2.3 Biological Transformations in Drinking Water Treatment

Biological transformations in drinking water treatment plants most commonly occur on granular filter media and/or in drinking water distribution systems. Significant biological activity can occur on fixed biofilms growing on the granular media. This growth is enhanced by the use of ozone which can increase the biodegradability of natural organic matter (Crittenden et al. 2005). Biological activity can also be prevalent in drinking water distributions systems, especially in those with long residence times.

Few studies have examined the biological degradation of transformation products in drinking water treatment plants or distribution systems (e.g, Huang and Banks 1996; Selim and Wang 1994), and more research is needed. Both Galluzzo et al. (1999) and Feakin et al. (1995) examined the formation of the transformation products deethylatrazine, deisopropylatrazine, and didealkylatrazine across a drinking water biofilter treating water containing atrazine. In general, minimal removal of the transformation products was observed. More discussion of the relative biodegradability of transformation products, as compared with that of the parents, is presented in the biological treatment section below.

2.4 Sorption to Coagulation Solids in Drinking Water Treatment

In coagulation/flocculation/sedimentation treatment of surface waters, a variety of coagulants is often used, including alum ($Al_2(SO_4)_3$), ferric iron salts ($FeCl_3$ or $Fe_2(SO_4)_3$), and polyaluminum chlorides (PACl). Each of these coagulants results in floc, with different characteristics, which coprecipitate with settled solids, or are removed on granular filters. In addition to the primary coagulant, polymers may also be added as coagulant aids (or as filter aids) to enhance flocculation and filtration by bridging colloidal material together. Because the combinations of coagulants and aids, as well as the colloidal solids in the source water, vary greatly from plant to plant, the surface chemistry of the settled solids (as well as materials coating the media in the granular filters) also varies greatly.

As discussed in previous chapters, transformation products often tend to have lower K_{OW} values than their corresponding parent compounds. This would then impart a greater water solubility, and a lesser tendency of transformation products to adsorb to these settled solids, than the parents. In general, very limited removal of transformation products on settled solids would be expected from coagulation processes in water treatment (Lykins et al. 1986; Adams et al. 2002; Jiang and Adams 2006).

2.5
Sorption to Activated Carbon in Drinking Water Treatment

2.5.1
Powdered Activated Carbon (PAC)

A key mechanism for removal of degradates in water treatment is via sorption to activated carbon. The two key forms of activated carbon commonly used in water treatment include powdered activated carbon (PAC) and granular activated carbon (GAC). PAC is used to remove taste-and-odor compounds (e.g, MIB and geosmin), natural organic matter (e.g., trihalomethane precursors), and synthetic organic chemicals. PAC can have a very high or low sorptive capacity for all of these compounds in a water treatment plant, depending on the dosage and type of PAC used.

Sorption to activated carbon occurs by various mechanisms including non-specific hydrophobic interactions, as well as electrostatic interactions between ionic functional groups of the sorbate and the activated carbon. As an example of the complexity of the sorption mechanism, the sorption of 18 pharmaceuticals on Acticarb AC800, at a dosage of 5 mg/l, is presented in Fig. 5 below (Westerhoff et al. 2005). Overall, there is no correlation ($\alpha = 0.05$) between the $\log K_{OW}$ and the percent adsorption for these compounds. However, if the pharmaceuticals that are ionic or have heterocyclic N are excluded, $\log K_{OW}$ does correlate well ($\alpha = 0.05$) with percent sorption on the PAC. This demonstrates that, for compounds for which ionic interactions are not important, the hydrophobicity/hydrophicity ratio (e.g., $\log K_{OW}$) may be a reasonable predictor of percent removal via non-specific sorption mechanisms.

PAC doses for taste-and-odor control in drinking water are typically on the order of 0.5–2 mg/l. These doses, however, are generally ineffective at removing appreciable amounts of many synthetic organic chemicals, such as the pesticides commonly found in treatment plants (Adams et al. 2002; Jiang and Adams 2006). For example, the dosage of Norit HDB PAC required

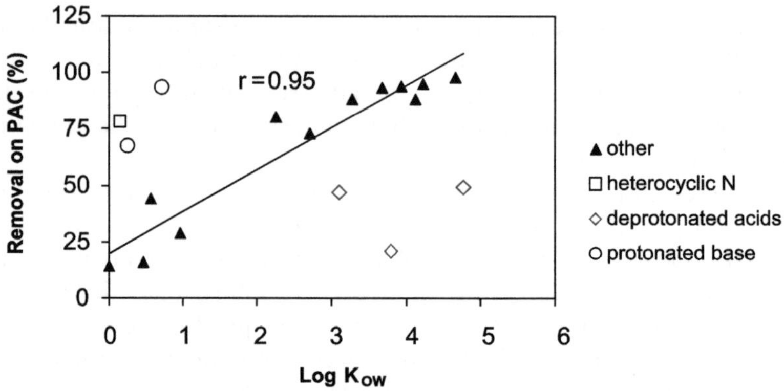

Fig. 5 Percent removal of 18 pharmaceuticals on 5 mg/l of PAC (Acticarb AC800). Note that sorption correlates well with the log octanol/water partition coefficient (log K_{OW}) in some cases, but not others where sorption mechanisms, other than non-specific hydrophobic/hydrophilic interactions, dominate (from Westerhoff et al. 2005)

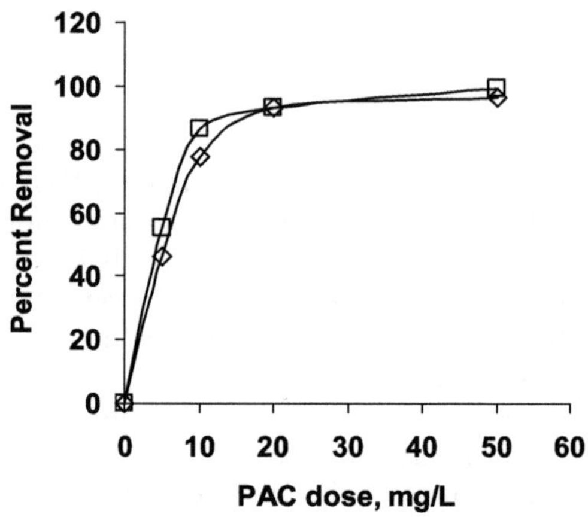

Fig. 6 Percent removal of didealkylatrazine from distilled water (DI) and from filtered Missouri River water using Norit HDB PAC. Initial concentrations of didealkylatrazine were 3 μg/l. Experiments were conducted at pH 7 for 4 h to reflect actual contact times achieved in a typical water treatment plant (from Jiang and Adams 2006)

to achieve 90% removal of didealkylatrazine (a transformation product of atrazine) from both lab and natural waters was 20 mg/l (Fig. 6) (Jiang and Adams 2006). Furthermore, approximately 50% removal was observed with a dose of 5 mg/l (which is still higher than that commonly used for taste-and-odor control).

Transformation products tend to have lower K_{OW} values, and, hence, would tend to be removed to an even lesser extent. As an example of the impact of these varied log K_{OW} values on the adsorption capacity of Norit HDB PAC,

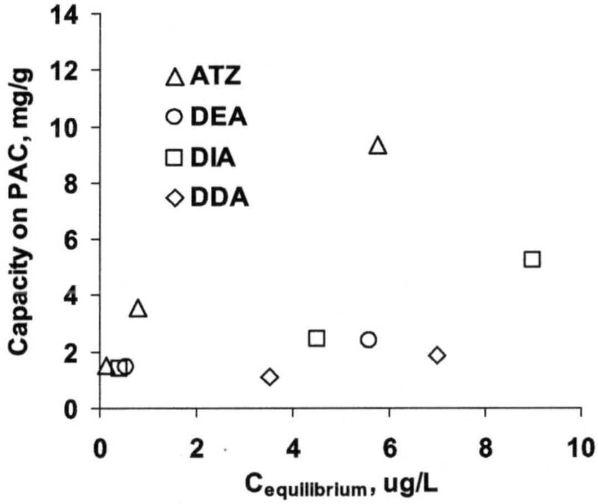

Fig. 7 Adsorption isotherms conducted individually on Norit HDB PAC in laboratory distilled water for atrazine (ATZ), deethylatrazine (DEA), deisopropylatrazine (DIA), and didealkylatrazine (DDA). Isotherms were conducted at pH 7 for 7 days with an initial concentration of chloro-*s*-triazine of 1 μg/l (from Jiang and Adams 2006)

we can examine adsorption data for atrazine (ATZ), deethylatrazine (DEA), deisopropylatrazine (DIA), and didealkylatrazine (DDA). Sorption data from Jiang and Adams (2006) show that the PAC has the greatest capacity for atrazine and the lowest capacity for didealkyatrazine. This reflects the strong correlation between K_{OW} values of 2.82, 1.78, 1.36, and 0.32 for atrazine, deethylatrazine, deisopropylatrazine, and didealkylatrazine, respectively, as calculated by the computational software KOWWIN (v. 1.65) (Fig. 7).

In a treatment study of chloro-*s*-triazine transformation products (and parent compounds), Jiang et al. (2006) found a wide range for removals for deethylatrazine and deisopropylatrazine in full-scale drinking water treatment plants. The study showed that treatment plants using PAC can vary from transformation product removal to nearly complete removal due to differences in type and dose of PAC, as well as water conditions.

2.5.2 Granular Activated Carbon (GAC)

GAC is most commonly used in water treatment plants as a replacement for anthracite on dual-media granular filters (Fig. 1). Alternatively, GAC may be used in post-filtration contactors. In either case, the GAC may be used for months or years before replacement is required (depending on treatment objectives, water quality, carbon type, and other factors).

Because transformation products may tend to sorb less than their parent compounds, understanding the relative advantages of GAC (as compared with PAC) is particularly important. GAC has an advantage over PAC in that GAC tends to come closer to equilibrium with the high influent concentration of a synthetic organic chemical, whereas PAC approaches equilibrium with the lower effluent (treated water) concentration. Therefore, for a physicochemically equivalent PAC and GAC, the GAC would achieve a much higher capacity for the synthetic organic chemical due to a greater aqueous-phase concentration.

Another factor is that insufficient contact time to approach equilibrium is generally provided in a drinking water treatment plant for PAC. Specifically, contact times of 2–4 h are generally provided, whereas, approaching equilibrium may take up to 4–5 d for PAC. On the other hand, GAC on GAC-capped filters or post-filter contactors does come nearly to equilibrium. For these reasons (i.e., more favorable equilibrium capacity, and closer approach to equilibrium), GAC has the potential to provide significantly more effective treatment for difficult to treat transformation products (as well as parent synthetic organic chemicals) in drinking water treatment plants.

GAC has two additional advantages over PAC. First, GAC is always in place on the filter or in a contactor, so that if an unexpected spike in an undesirable synthetic organic chemical (e.g., from a spill) enters the plant, the activated carbon is already treating the water at its full capacity. Second, the adsorbed transformation products will be thermally destroyed (or volatilized) during the GAC reactivation process. PAC, on the other hand, is not regenerated, but is retained with the backwash water. Therefore, the transformation product-laden PAC in backwash solids may pose a significant risk of leaching degradates from their final location in a lagoon or landfill into underlying groundwaters.

An example of the effectiveness of GAC for transformation product control is the nearly complete removal at pilot-scale of deethylatrazine and deisopropylatrazine on both a GAC-capped filter and a GAC-contactor (Hulsey et al. 1993). This is only comparable with the partial removal of these same compounds using PAC at dosages of 5 mg/l (or less), as demonstrated in Fig. 5.

A study by Sutherland et al. (2005) examined treatability using GAC of fuel oxygenates, including *t*-butanol (TBA), an advanced oxidation transformation product of methyl-*t*-butyl ether (MTBE). The study showed that, while MTBE is partially treatable by GAC, the transformation product, TBA, is completely non-adsorbable and non-treatable with GAC.

2.6 Membranes in Drinking Water Treatment

Membrane operations in water treatment processes include (in order of decreasing pore size): microfiltration, ultrafiltration, nanofiltration, and reverse osmosis. In general, microfiltration and ultrafiltration would have little effect on the removal of transformation products due to their relatively large pore size, unless the transformation products were sorbed to solids (e.g., colloidal material, powdered activated carbon, etc.) that were effectively retained on the micro- or ultra-filters. Both nanofiltration and reverse osmosis, however, would be expected to be relatively effective at removing many transformation products, especially those of higher molecular weight. While few studies have been conducted on membrane treatment of transformation products, the technology can be expected to be nearly as effective for transformation products as for the parent synthetic organic chemicals (though differences in molecular weight, molecular diameter, and $\log K_{OW}$ could cause a significantly different rejection of transformation products, as compared with parents).

A key issue with the use of membranes for treatment of transformation products of synthetic organic chemicals is that the pollutants are simply concentrated into a reject stream that must be dealt with in some manner. The treatment and disposal of this waste stream are much more problematic when the waste contains toxic compounds (e.g., some transformation products) rather than, for example, simply natural organic matter.

3 Fate of Transformation Products in Wastewater Treatment

3.1 Wastewater Treatment Processes and Operation

A typical process that is used to treat municipal wastewater is shown in Fig. 8. There are many similarities, with respect to the fate of transformation byproducts, between wastewater treatment and drinking water treatment in that abiotic transformation, biotic transformation, and partitioning reactions (especially sorption) may occur. Key differences, however, are that biological reactions during wastewater treatment are much more robust and effective than those during drinking water treatment. Also, the solids concentrations are much higher, thereby, providing a greater potential for sorption. These processes are discussed in more detail below.

3.2 Abiotic Transformations in Wastewater Treatment

3.2.1 Chemical Oxidation in Wastewater Treatment

The primary manner for adding an oxidant during municipal wastewater treatment is with chlorine disinfection of the secondary effluent, after biological treatment (Fig. 8). Chlorine disinfection is commonly used and effective

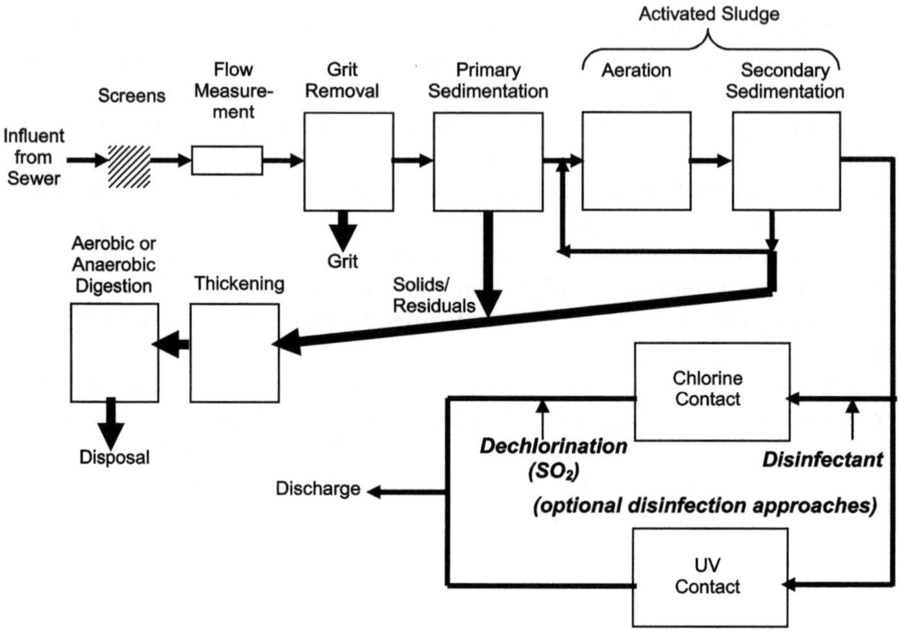

Fig. 8 A typical treatment scheme for municipal wastewater, using conventional activated sludge. Other secondary treatment processes are also common, including trickling filters and oxidation ditches

in reducing the release of pathogens into a stream or river receiving the treated wastewater. After chlorine addition, contact time must be provided using a contact basin.

Chemical oxidation has the potential to both form transformation products and to remove them during wastewater treatment. However, very little work has been conducted on the chemical oxidation of transformation products (or parent synthetic organic chemicals) during wastewater treatment disinfection. Work by Qiang et al. (2006) on the oxidation of sulfonamide antibiotics in an anaerobic lagoon effluent treating swine manure showed that high chlorine dosages (on the order of 500 mg/l) were needed to significantly remove these antibiotics from the wastewater.

On the other hand, Huber et al. (2005) found that low dosages of ozone can be effective at removing many pharmaceuticals from municipal wastewater. By extrapolation, ozone may be more or less able to react with, and transform, transformation products in wastewater depending on the reactivity of a particular compound with ozone and its concentrations. Furthermore, as pH increases from nominally 6 to 10, an increasingly larger fraction of ozone decomposes via hydroxide-catalyzed reactions to hydroxyl radicals, which are highly reactive with many compounds (including transformation products and parents, as well as background constituents and scavengers). Thus, pH can play an important role in determining the extent of direct or indirect reactions of ozone with transformation products.

Because ozone, chlorine, and UV are all common disinfectants in both drinking water and wastewater treatment, the formation and subsequent conversion of transformation products in wastewater can be similar to those observed in drinking water (as discussed above). However, much higher disinfectant or oxidant dosage would typically be required in wastewater due to the scavenging of the oxidant (or UV) by the much more concentrated background constituents in wastewater as compared to drinking water.

3.2.2 Chemical Reduction in Wastewater Treatment

Following the chlorine chemical contact for disinfection, described in the previous section, residual chlorine must be removed using a dechlorination agent such as sulfur dioxide (Fig. 8). This quenching of the chlorine (or chloramine) by a strong reductant provides a potential for the chemical reduction of transformation products in the wastewater. Minimal research has addressed these reductive reactions for transformation products of synthetic organic chemicals.

Further opportunity for the chemical reduction of transformation products exists in various locations within the wastewater treatment plants in zones where the oxidation reduction potential (ORP) decreases to a negative (reducing) range. Due to the rapid depletion of oxygen from the wastewater, this may occur in any area of a sewer or treatment plant that is not adequately oxygenated. Additionally, transformation products sorbed to primary solids or secondary (biological) solids are often treated in anaerobic digesters where abiotic reductive reactions (as well as anaerobic biodegradation) may occur.

3.2.3 Hydrolysis in Wastewater Treatment

In general, the pH range in a wastewater treatment plant is not likely to promote hydrolysis reactions to a great extent beyond that already occurring in the raw sewage. However, hydrolysis can occur, even at a neutral pH and, therefore, provide potential for hydrolyzing transformation products in a wastewater treatment plant.

3.2.4 Photolysis in Wastewater Treatment

Ultraviolet (UV) disinfection is a commonly-used alternative wastewater treatment to chlorine disinfection. While few studies have addressed photolysis of transformation products in wastewater, it is logical to assume that much greater UV absorbance by background compounds will be present in wastewater as compared to drinking water. Because the concentration of most transformation products will be many orders of magnitude lower than the concentration of the background material, photolysis would not be expected to be a significant reaction for most transformation products.

3.3 Biological Transformations in Wastewater Treatment

Biological degradation (and formation) of transformation products in secondary (biological) treatment can typically be expected to follow traditional Monod kinetics

$$\frac{\mathrm{dTP}}{\mathrm{d}t} = -\frac{U_{\mathrm{max}}[\mathrm{MLSS}][\mathrm{TP}]}{Y(K_{\mathrm{C}} + [\mathrm{TP}])},$$

where [TP] is the concentration of the transformation product (mol/l), [MLSS] is the mixed liquor suspended solids (or biomass) concentration, U_{max} is microorganism's maximum growth rate, K_{C} is the half-maximum growth rate, and Y is a yield coefficient (see Grady et al. 1999). When a compound being degraded is at a relatively low concentration (such as with most transformation products of synthetic organic chemicals), the Monod equation reduces to a pseudo-first order equation

$$\begin{aligned}\frac{\mathrm{dTP}}{\mathrm{d}t} &= -\frac{U_{\mathrm{max}}[\mathrm{MLSS}][\mathrm{TP}]}{Y(K_{\mathrm{C}} + [\mathrm{TP}])}\\ &= -\frac{U_{\mathrm{max}}[\mathrm{MLSS}][\mathrm{TP}]}{Y(K_{\mathrm{C}} + [\cancel{\mathrm{TP}}])}\\ &= -\left(\frac{U_{\mathrm{max}}}{YK_C}\right)[\mathrm{MLSS}][\mathrm{TP}]\\ &= -k'[\mathrm{MLSS}][\mathrm{TP}]\end{aligned}$$

where k' is the pseudo-first-order biodegradation rate constant. The MLSS concentration is relatively high (e.g., 1500–4000 mg/l) in activated sludge, and the biosolids concentration is even higher in concentrated sludge or biofilms (e.g., waste activated sludge, trickling filters, etc.), thereby promoting enhanced biodegradation rates. Thus, a wide range of biodegradabilities for transformation products can be expected for wastewater treatment plants ranging from readily biodegradable to biorecalcitrant.

A look at several examples is instructive. Stefan and Bolton (1998) determined that advanced oxidation of 1,4-dioxane leads to the transformation products of mono- and di-formate esters of 1,2-ethanediol, various organic acids (e.g., formic, acetic, glycolic, and oxalic acids), and aldehydes (e.g., formaldehyde, and acetaldehyde). Adams et al. (1994) had previously studied the biodegradability of these mixtures of 1,4-dioxane transformation products and determined that they were significantly more biodegradable than the parent compound, 1,4-dioxane.

Adams et al. (1996) studied the biodegadability of hydroxyl-radical mediated transformation products of nonionic surfactants. The transformation products of linear secondary alcohol ethoxylates and ethylene oxide/propylene oxide surfactants were observed to be, on average, readily biodegradable as opposed to more biorecalcitrant parent compounds. On the other hand, the initial (lower oxidant dosage) transformation products of alkylphenol ethoxylates, (powerful endocrine disrupting chemicals), were less biodegradable than the parent compounds. These compounds were hypothesized to be alkylphenol ethoxylates with shorter EO chains. Further oxidation of the initial transformation products were more readily biodegradable (and were hypothesized to have ring cleavage) (Adams et al., 1996). Similar results were observed by Kitis et al. (2000) in a subsequent, and more detailed study.

Another study by Dantas et al. (2007) determined that the ozone transformation products of bezafibrate were much more readily biodegradable than the parent compounds. The toxicity of the degradates was also less than that of the parent compound.

In another study, the biodegradability of the ozonation transformation products of chloro-, nitro- and amino-phenols were studied (Adams et al., 1997). The study determined that the transformation products of chloro- and nitro-phenols were more readily biodegradable than the parent compounds. Transformation products of amino-phenols were, however, significantly more biorecalcitrant than those of the parent.

Another example of the differences in biodegradability of advanced oxidation (hydroxyl-radical mediated) transformation products are for quaternary amines. A study by Adams and Kuzhikannil (2000) showed that transformation products of alkyldimethylbenzyl ammonium chlorides (Barquats) were, on average, significantly more biodegradable than the parent compounds. However, the biodegradability of transformation products of dioctyl-dimethyl ammonium chlorides (Bardac LF) was significantly less than that of the parent.

These studies demonstrate that generalizations regarding the relative biodegradability of transformation products as compared to the parent compounds, are difficult, and should be carefully made. Preferably, treatability studies should be conducted to assess biodegradability for any given system.

3.4 Sorption to Settled Primary and Secondary (Biological) Solids in Wastewater Treatment

Significant sorption of transformation products to settled primary solids and secondary (or biological) solids may often be a primary loss route in wastewater treatment, especially for transformation products with higher $\log K_{\mathrm{OW}}$ values. The major mechanism of sorption is often due to the partitioning of the synthetic organic chemical into the organic carbon phase of the settled solids. However, other mechanisms are also possible, including cation exchange, anion exchange, and chemisorption (or formation of covalent bonds).

For compounds at very low relative concentrations, linear adsorption isotherms to solids may be dominant, i.e.,

$$K_{\mathrm{D}} = \frac{[\text{sorbed TP}]}{[\text{aqueous TP}]},$$

where K_{D} is the linear isotherm coefficient. If partitioning of a transformation product is into the organic fraction of the solid phase, the K_{D} term becomes

$$\begin{aligned} K_{\mathrm{D}} &= \frac{[\text{sorbed TP}]}{[\text{aqueous TP}]} \\ &= \frac{[\mathrm{TP}]_{\text{sorbed to organic matter}} f_{\text{organic fraction of solid}}}{[\mathrm{TP}]_{\text{aqueous}}} \\ &= K_{\mathrm{OM}} f_{\text{organic fraction of solid}}, \end{aligned}$$

where K_{OM} is the partition coefficient onto organic matter, and $f_{\text{organic fraction of solid}}$ is the fraction by mass of the solids comprised by this organic matter.

K_{D} is unique to any particular system and is difficult to predict with accuracy. Factors, including the nature of the transformation product, pH (as it relates to speciation and ionization), temperature, solids concentrations, background organics, total dissolved solids, and many other factors, can play a role in affecting the K_{D}.

A related parameter is the octanol/water partition coefficient (K_{OW}) which, for a transformation product, is related to the partitioning of a transformation product between octanol and water in a "clean" system. Thus, K_{OW} is a specific K_{OM} for partitioning into one specific organic material (i.e., octanol). Thus, when sorption into the organic phase of wastewater solids is the dominant sorption mechanism, then K_{OW} is often closely related to K_{D} for the transformation product and the biosolids. Log K_{OW} values can be analyzed experimentally, and can also be estimated using computational techniques such as EPIWIN software (US EPA 2007). When a transformation product has a lower log K_{OW}, it is generally going to sorb less to biosolids (unless other mechanisms such as ion exchange dominate). For oxidation products, the log K_{OW} of transformation products will often be lower than that of the parent compounds due to the addition of anionic carboxylic, hydroxyl, or phenol structure to the parent. Additionally, because these groups are ionizable, pH may have a significant effect on the log K_{OW} for a transformation product and, hence, its degree of sorption to biosolids.

4 Summary

Transformation products of synthetic organic chemicals may be more, equally, or less toxic to humans (and the environment) than the parent compounds they arise from. Concentrations of transformation products may be significantly higher than the parent compounds in natural and/or engineered systems. Thus, the human health risk associated with toxicity and exposure has the potential to be greater for some transformation products than for the parent compounds.

While a majority of fate and effects research has focused on parent compounds, there has also been an emerging realization that equal focus may need to be placed on understanding the role of transformation and partitioning mechanisms on the formation and removal of transformation products in drinking water and wastewater treatment. While specific experimental studies of the plethora of transformation products are insufficient, some generalizations regarding transformation and partitioning of transformation products can be made based on parameters specific to the transformation product such as K_{OW} and chemical oxidation rate constants.

References

1. Acero J, Stemmler K, von Gunten U (2000) Environ Sci Technol 34:591
2. Acero J, Haderlein S, Schmidt T, Suter M, Von Gunten U (2001) Environ Sci Technol 35:4252
3. Adams C, Ma Y, Shi H, Chamberlain E, Wang T (2007) (ongoing unpublished research)
4. Adams C, Cozzens R, Kim B (1997) Water Res 31:2655
5. Adams C, Kuzhikannil J (2000) Water Res 34:668
6. Adams C, Wang Y, Loftin K, Meyer M (2002) J Environ Eng 128:253
7. Adams C, Randtke S (1992) Environ Sci Technol 26:2218
8. Adams C, Scanlan P, Secrist N (1994) Environ Sci Technol 28:1812
9. Adams C, Spitzer S, Cowan R (1996) J Environ Eng 122:477
10. Adams C, Thurman E M (1991) J Environ Qual 20:540
11. Adams C, Wang Y, Loftin K, Meyer M (2002) J Environ Eng 128:253
12. Buth J, Arnold W, McNeill K (2007) Environ Sci Technol 41:6228

13. Crittenden J, Trussell R, Hand D, Howe K, Tchobanoglous G (2005) Water Treatment: Principles and Design 2nd edn. Wiley, Hoboken, NJ
14. Dantas R, Canterino M, Marotta R, Sans C, Esplugas S, Andreozzi R (2007) Water Res 41:2525
15. Doll T, Frimmel F (2004) Water Res 38:955
16. Feakin S, Bubbins B, McGhee I, Shaw L, Burns R (1995) Water Res 19:1681
17. Galluzzo M, Banerji S, Bajpai R, Surampalli R (1999) Pract Period Hazard. Toxic Radioact Waste Manag 3:163
18. Grady C, Daigger G, Lim H (1999) Biol Wastewater Treat, 2nd edn. Marcel Dekker, New York
19. Huang C, Banks M (1996) J Environ Sci Health B 31:1253
20. Huber M, Göbel A, Joss A, Hermann N, Löffler D, McArdell CS, Ried A, Siegrist H, Ternes T, Von Gunten U (2005) Environ Sci Technol 39:4290
21. Hulsey RA, Randtke SJ, Adams C, Long BW (1993) Ozone Sci Eng 15:227
22. Jiang H, Adams C (2006) Water Res 40:1657
23. Jiang H, Adams C, Graziano N, Roberson A, McGuire M, Frey M (2006) Environ Sci Technol 40:3609
24. Kitis M, Adams C, Kuzhikannil J, Daigger G (2000) Environ Sci Technol 34:2561
25. Lau T, Chu W, Graham N (2007) Water Res 41:765
26. Lopez A, Mascolo G, Tiravanti G, Passino R (1998) J Anal 53:856
27. Lykins B, Koffskey W, Miller R (1986) J Am Water Works Assoc 78:66
28. Mascolo G, Lopez A, James H, Fielding M (2001a) Water Res 35:1695
29. Mascolo G, Lopez A, James H, Fielding M (2001b) Water Res 35:1705
30. Qiang Z, MacCauley J, Mormile M, Surampalli R, Adams C (2006) J Agric Food Chem 54:8144
31. Selim M, Wang J (1994) Environ Toxicol Chem 13:3
32. Sharpless C, Siddiqui M, Atasi K, Linden K (2003) Proc Water Quality Technology Conference, pp 763–775
33. Sinclair C, Boxall A, Parsons S, Thomas M (2006) Environ Sci Technol 40:7283
34. Stefan M, Bolton J (1998) Environ Sci Technol 32:1588
35. Sutherland J, Adams C, Kekobad J (2005) J Environ Eng 131:623
36. US Environmental Protection Agency (2007) EPI Suite v3.20. US EPA, 2007
37. Vogna D, Marotta R, Napolitano A, Andreozzi R, D'Ischia M (2004) Water Res 38:414
38. Westerhoff P, Yoon Y, Snyder S, Wert S (2005) J Environ Eng 39:6649

Part III
Effects of Transformation Products

Hdb Env Chem Vol. 2, Part P (2009): 177–204
DOI 10.1007/698_2_019

Published online: 14 August 2009

Ecotoxicity of Transformation Products

Chris J. Sinclair[1] (✉) · Alistair B.A. Boxall[2]

[1]The Food and Environment Research Agency, Sand Hutton, York YO41 1LZ, UK
chris.sinclair@fera.gsi.gov.uk

[2]University of York, Heslington, York YO10 5DD, UK

Abstract While a large body of information is available on the environmental effects of parent chemicals, we know much less about the effects of transformation products. However, transformation products may be more toxic, more persistent and more mobile than their parent compound. An understanding of the ecotoxicity of transformation products is therefore essential if we are to accurately assess the environmental risks of synthetic chemicals. This chapter therefore uses data on pesticides and their transformation products to explore the relationships between parent and transformation product ecotoxicity to aquatic and terrestrial organisms and describes the potential reasons why a transformation product may be more toxic than its parent compound. As it is not feasible to experimentally assess the ecotoxicity of each and every transformation product, this chapter also describes and evaluates the use of expert systems, read-across methods and quantitative structure activity relationships for estimating transformation product ecotoxicity based on chemical structure. Finally, experimental and predicted ecotoxicity data are used alongside monitoring data for parent pesticides and their transformation products to illustrate how the risks of parent and transformation product mixtures can be assessed.

Keywords Ecotoxicity · Expert systems · Mixture · Mode of action · QSAR · Read-across · Toxicity

Abbreviations

EC_{50} Median effective concentration
EFSA European Food Safety Authority
EPA Environmental Protection Agency
EU European Union
HSE Health and Safety Executive
LC_{50} Concentration causing 50% mortality
$\log K_{ow}$ Octanol/water partition coefficient
OECD Organisation for Economic Cooperation and Development
OP Organophosphorus insecticide
pK_a Acid dissociation constant
PNEC Predicted no effect concentration
PSD Pesticide Safety Directorate
QSAR Quantitative structure–activity relationship
RQ Risk quotient

1 Introduction

The detrimental impact of transformation products to environmental and human health is not a recent issue. Some of the most publicized historical cases of environmental and human health effects of pesticides have been shown to be due to transformation products rather than the parent compounds. For example, the chronic egg shell thinning in raptors and fish-eating birds (e.g., peregrine falcon and brown pelican) was primarily due to 1,1-dichloro-2,2-bis(4-chlorophenyl)ethylene (DDE), a transformation product of the organochlorine insecticide DDT [1], whilst the mammalian carcinogenic effects of the plant growth regulator daminozide were due to the transformation product unsymmetrical dimethylhydrazine [2]. The insecticide DDT went from hero to villain in a relatively short space of time. Paul Müller won the 1948 Nobel Prize in Physiology and Medicine for identifying the potent effect of DDT on arthropods [3], whilst only fifteen years later it was suggested that organochlorine insecticides were drastically effecting bird populations [4]. Nowadays it would be improbable that highly lipophilic pesticides such as DDT, that breakdown to form persistent lipophilic and highly toxic transformation products would be developed and be capable of passing the rigorous risk assessment procedures [5]. However it is important that the impact of transformation products formed from any chemical intentionally or unintentionally released to the environment are considered.

Pesticides are a highly regulated group of chemicals and undergo a very detailed risk assessment process which, depending on the specific pesticide and its subsequent degradation, can include detailed information on some of its transformation products. An evaluation of the ecotoxicological and toxicological potential of these compounds can however be hampered by a lack of available data. Almost half the transformation products considered during the aquatic risk assessment of Belfroid et al. [6] which considered 20 pesticides from a variety of chemical classes had no available ecotoxicological data at all. Moreover only a quarter of those transformation products under investigation had acute aquatic ecotoxicological data for fish, daphnids and algae of a quality considered "moderate to sufficient" to proceed with the estimation of risks. Similarly in another study investigating transformation product aquatic acute ecotoxicity, less than 20% of the identified 485 environmental transformation products from 60 pesticides had available data [7]. The data collation for both of these studies was primarily within the publicly available literature, however, regulatory data specific to pesticides (and their transformation products) have now become widely available from regulatory authorities from various geographic regions, e.g., Reregistration Eligibility Decision documents from the US Environmental Protection Agency (EPA) [8], Review Reports from the European Union (EU) [9, 10] and Evaluation Documents from the Pesticide Safety Directorate (PSD) in the UK [11] so more data is now becoming available. Unfortunately, similar data resources are not available for parent compounds from other product types and hence the risks of transformation products arising from industrial chemicals, personal care products, biocides and human and veterinary pharmaceuticals are less well understood [12].

It is generally perceived that the toxicological effects of transformation products are unintentional impacts that need to be managed if the benefits of the parent compound is to be obtained. However the desired toxic effect for some pesticides is actually exhibited by a transformation product rather than the applied parent compound. These pesticides, often prefixed with "pro", are designed to work in this manner to increase the desired efficacy. This prefix was first applied to human pharmaceuticals where it was used to describe the requirement for some compounds to undergo structural modification before they become active and has subsequently been applied to pesticides [13]. Generally pesticides and in particular some groups of insecticides are designed to work in this manner to increase the quantities that reach the target site. For example, the organophosphorus insecticides act on the enzyme acetylcholine esterase that hydrolyzes the transmitter acetylcholine present at a nerve synapse [5]. A number of these insecticides undergo metabolic activation via oxidative desulphurization (i.e. the P=S moiety to a P=O moiety) directly in living organisms, which results in the formation of much more potent acetylcholine esterase inhibitors (e.g., diazinon to diazoxon) [14]. The use of precursors rather than the highly active molecules can increase pesticide efficacy due to better insect cuticle penetrability and increased environmental residence time [15]. Generally this concept is considered during risk assessment which would normally include the biologically active molecule as well as the compound applied into the environment.

In this chapter, we use data on the ecotoxicity of pesticides and their associated transformation products to illustrate the relationships between parent and transformation product toxicity; we describe the possible reasons why a transformation product may be more ecotoxic than its parent compound and explore the use of predictive models for estimating the ecotoxicity of transformation products in the absence of data. Finally we discuss the concept of parent compound and metabolite interactions and discuss how the risks of these mixture, which will occur in the natural environment, can be assessed.

2
Comparison of Parent and Transformation Product Ecotoxicity

As discussed earlier extensive data are available on the ecotoxicological effects of pesticide transformation products whilst only limited ecotoxicological data are available on the ecotoxicity of transformation products of other classes of compound such as human and veterinary pharmaceuticals, industrial chemicals and biocides. Generally transformation products are considered to be less toxic to non-target organisms, however some exhibit an equivalent or increased potency when compared to the parent compound. Therefore, in the following sections we take data on the aquatic (daphnid) and terrestrial (earthworm) ecotoxicity of pesticide transformation products and compare this to data on the ecotoxicity of the associated parent pesticide in order to explore the relationships between parent and transformation product ecotoxicity in aquatic and terrestrial systems.

2.1
Effects of Transformation Products on Daphnids

In order to explore the relationships between parent and transformation product ecotoxicity to invertebrates, data relating to the acute aquatic ecotoxicity of transformation products and their respective pesticides to the water flea *Daphnia* sp. were collated. The majority of data points were for the species *Daphnia magna* whilst some data were for either *Daphnia pulex* or undefined daphnid species, all these data were treated as comparable. Data collection focused on the end-point stipulated in the OECD guideline, 48 h EC_{50} (immobilisation) [16]. Data collection principally focused on pesticides evaluation documents and were supplemented with data collated for the EU SEEM project [17]. Where multiple values were identified for a pesticide or transformation product a geometric mean value was used. Where transformation product ecotoxicity data were identified in the evaluation documents with no respective pesticide data, alternative data sources were used to provide a comparison [18, 19]. Initially 255 pesticide/transformation product data comparisons were identified which comprised 120 pesticides and 245 transformation products. However, 36 comparisons were removed because the data was not suitable for comparison purposes since both data points were represented by "greater than values" and therefore impossible to determine which compound was the most potent. Some of the comparisons were represented by inequality data with the comparable value a specific numeric. This occurred 108 times for transformation products and 23 times for pesticides. During data comparisons 33 points were removed from these data sets because it was impossible to identify the most potent compound. Therefore, including the data where both compounds were represented by numerics, 186 data comparisons between pesticides and their transformation products were available to examine the relationships between their acute aquatic ecotoxicity to daphnids. A further 29 data comparisons were removed from the analysis because a molecular structure was not available for 27 transformation products. Structure was important to the analysis as the molecular weight was needed so that comparisons between parent and transformation product could be undertaken in mmol/L and so that selected physico-chemical properties could be estimated. A comparison of the transformation product and parent data is provided in Fig. 1.

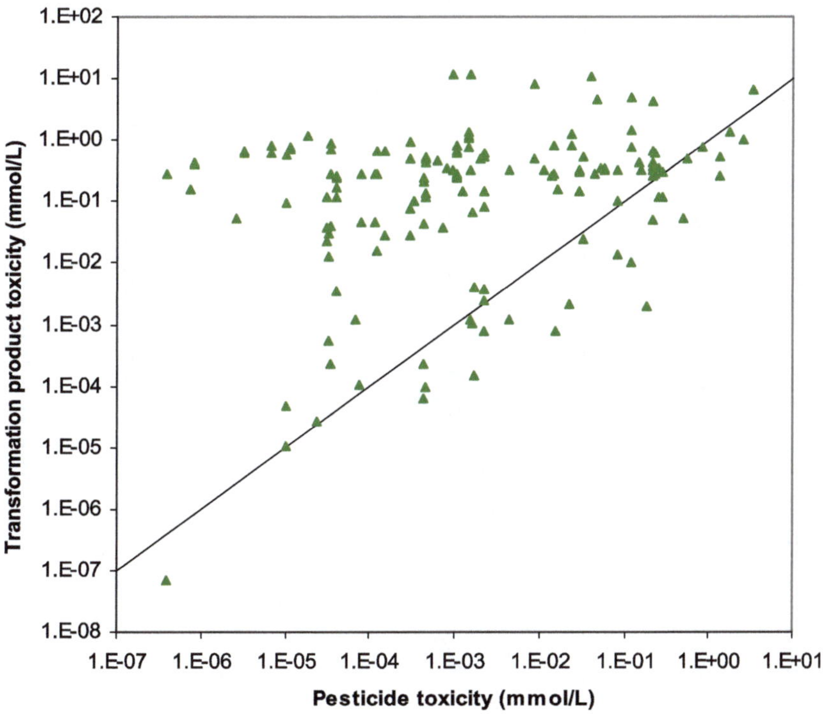

Fig. 1 A comparison of the acute toxicity of pesticides and their respective transformation products to daphnids. The *solid line* represents equal toxicity ($x = y$)

Table 1 Transformation products that exhibit a potency to daphnids greater than one order of magnitude more than their parental pesticides

Pesticide Name	Structure	Daphnid toxicity (mmol/L)	Pesticidal class	Log K_{ow} [a]	pK_a [a]	Transformation product Name	Structure	Daphnid toxicity (mmol/L)	Log K_{ow} [b]	pK_a [c]	Possible explanations for increase in potency
Bifen-azate	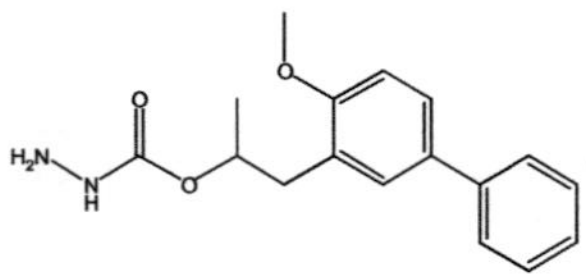	0.002	Acaricide	3.4	12.94	D3598	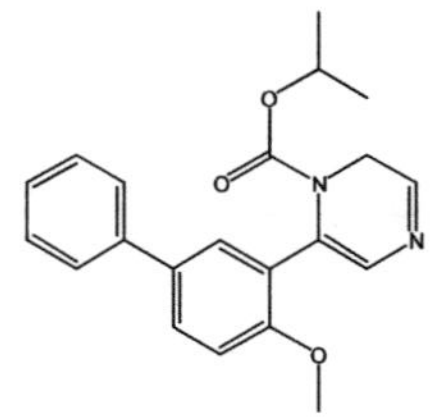	0.0001	4.08	8.48	Increase in hydro-phobicity
Thio-phanate-methyl	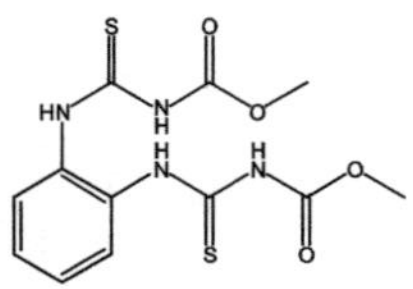	0.016	Fungicide	1.5	7.28	Car-bend-azim	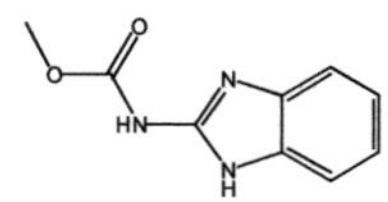	0.0008	1.42	7.08	Pro-pesticide
Acephate	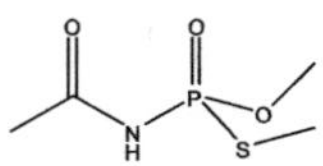	0.187	Insecticide	– 0.89	Not appli-cable	Meth-amido-phos	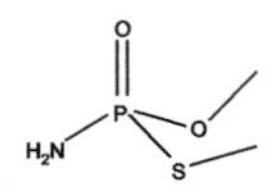	0.0019	– 0.91	Not appli-cable	Pro-pesticide

Table 1 (continued)

Pesticide Name	Pesticide Structure	Daphnid toxicity (mmol/L)	Pesticidal class	Log K_{OW} [a]	pK_a [a]	Transformation product Name	Transformation product Structure	Daphnid toxicity (mmol/L)	Log K_{OW} [b]	pK_a [c]	Possible explanations for increase in potency
Propyz-amide	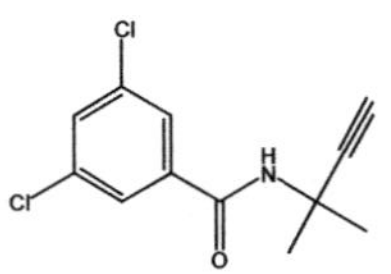	0.022	Herbicide	3.3	Not appli-cable	RH-24644	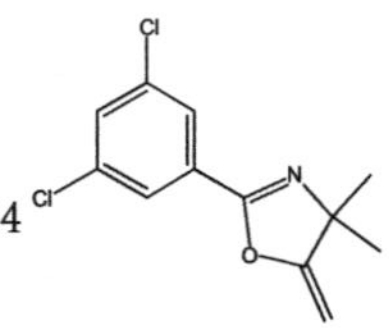	0.0021	4.83	4.68	Increase in hydro-phobicity
Amido-sulfu-ron	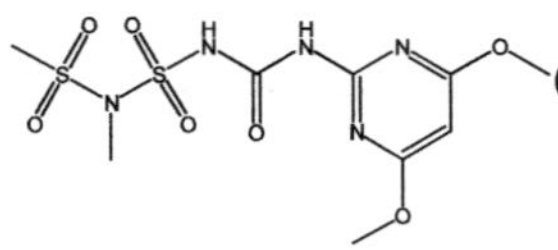	0.123	Herbicide	1.63	3.76	Hoe 101630	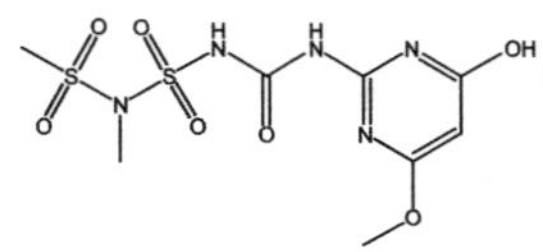	0.0101	0.46	0.43, 1.08, 1.72	?

[a] Hydrophobicity and dissociation data from e-Pesticide Manual [20].
[b] Transformation product hydrophobicity calculated from the mean of $C\log P$ (Biobyte Corp.), KOWWIN (USEPA) and AlogPS (VCCLAB)
[c] Dissociation calculated using SPARC (University of Georgia)

The majority (83.3%) of transformation products exhibited a toxicity to daphnids equal to or lower than their respective parental pesticide (Fig. 1). In total, 26 transformation products exhibited an increase in toxicity when compared to the parental pesticide, 54% of these, transformation products were produced from pesticides that control plants (as herbicides or plant growth regulators) and 31% had parents that act on animal pests (i.e., as an insecticide, acaricide, and/or nematicide) with the remaining having fungicidal parents. The majority of herbicidal transformation products that exhibited an increase in toxicity to daphnids compared to the parent pesticide resulted from the breakdown of sulfonylurea herbicides. Five transformation products (3.2%) exhibited a potency to daphnids which was at least an order of magnitude greater than parental pesticide value whilst none exhibited a potency greater than two orders of magnitude (Table 1). These increases in potency are probably attributed to increases in hydrophobicity of the transformation product compared to the parent compound or the transformation product representing the biological active component of the pesticide, i.e. the pesticide is a pro-pesticide.

These observations are similar to previous studies that compared the ecotoxicity of pesticide parent compounds and their transformation products to fish, daphnids and algae [7]. In this study, increases in toxicity were attributed in changes in hydrophobicity or dissociation behaviour between the parent and the transformation product, the fact that the parent was a pro-pesticide, the transformation product maintaining the toxicophore of the parent compound or the transformation product gaining a different and more potent mechanism of action. These mechanisms for increases in toxicity are described in more detail later.

2.2 Effects of Transformation Products on Earthworms

To explore the relationships between parent and transformation product ecotoxicity to earthworms, data were collated on the acute ecotoxicity to earthworms for pesticides and their transformation products. These were acute, generally 14 d, LC_{50} data and were collected from pesticide evaluation documents of the UK PSD and Health and Safety Executive (HSE) [11], US EPA [8], EU [9, 10], the Canadian Pest Management Regulatory Agency (PMRA) [21] and the European Food Safety Authority (EFSA) [22]. Collated data were supplemented by data collated by the EU SEEM project [17]. Ultimately this provided 142 comparisons between pesticides and their transformation products.

Earthworm toxicity data are often reported, not as specific numbers but as an inequality, e.g. $> 1000\,\mathrm{mg\,kg^{-1}}$ soil, making it very difficult to undertake a straightforward correlation as used in the previous section for

Table 2 Hazard classification system for acute toxicity for soil dwelling organisms (from [23])

$mg\,kg^{-1}$ soil/substrate	≤ 50	> 50250	$> 250 \leq 1000$	> 1000
Classification	Very toxic	Toxic	Harmful	Not classified

daphnids, therefore data have been analysed in a qualitative manner, using the terrestrial hazard classification proposed by Jensen [23] (Table 2). This classification system uses the phrase "not classified" to indicate a low potency to earthworms, i.e. $> 1000\,mg\,kg^{-1}$ soil, to "very toxic" $< 50\,mg\,kg^{-1}$ soil.

Using this system, five transformation products were classified as very toxic to earthworms, carbendazim, methomyl, 3,5,6-trichloro-2-pyridinol, 3-methyl-4-nitrophenol and 4-methoxybiphenyl. Carbendazim and methomyl are commercially marketed fungicides and insecticides in their own right. The transformation products 3,5,6-trichloro-2-pyridinol and 3-methyl-4-nitrophenol are transformation products from the organophosphorus (OP) insecticides chlorpyrifos and fenitrothion respectively. The molecular structure of these two transformation products do not contain the moiety of the parent compound considered to elicit the acteylcholinestearse inhibition exhibited by the parent pesticides [5]. Whilst 4-methoxybiphenyl is a transformation product of the acaricide bifenazate.

Twenty-four pesticides and 21 transformation products had earthworm toxicity data that did not easily lend itself to the hazard classification system because it is impossible to place the data in one of the proposed categories (e.g., > 10 mg/kg soil). Therefore 101 comparisons could be made between earthworm toxicity of transformation products and their respective parental pesticide (Table 3).

The majority (91%), of transformation products were allocated an equal or lower hazard classification than their parent pesticides (Table 3), with

Table 3 Comparison of the hazard classification relationships for acute earthworm toxicity between pesticides and their transformation products (occasions were transformation product toxicity is greater than pesticide toxicity are in bold)

		Transformation products			
	Classification	Not classified	Harmful	Toxic	Very toxic
Pesticides	Not classified	32	**4**	**4**	–
	Harmful	12	3	–	–
	Toxic	16	3	–	**2**
	Very toxic	12	2	2	2

Table 4 Transformation products that were classified in a higher risk category for earthworms than their parental pesticides

Pesticide Name	Pesticide Structure	Earthworm toxicity (mg/kg soil)	Hazard classification	Log K_{ow}^{a}	pK_a[a]	Transformation product Name	Transformation product Structure	Earthworm toxicity (mg/kg soil)	Hazard classification	Log K_{ow}^{b}	pK_a[c]
Chlorotoluron	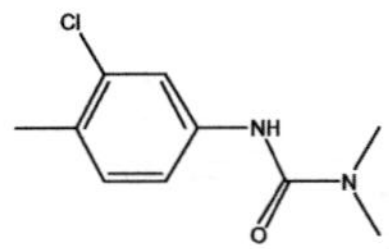	> 1000	Not classified	2.5	Not applicable	3-(3-chloro-*p*-tolyl)-1-methylurea	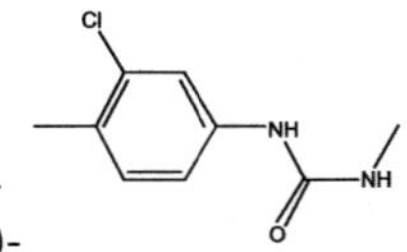	697	Harmful	2.42	0.36
Methiocarb	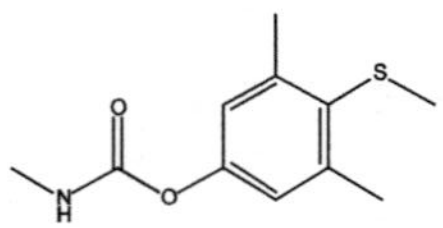	1322	Not classified	3.18	Not applicable	Methiocarb methoxy sulfone	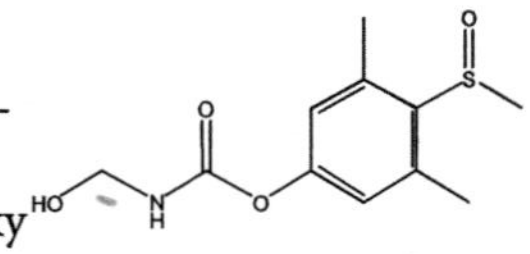	562	Harmful	0.19	8.49
Prosulfuron	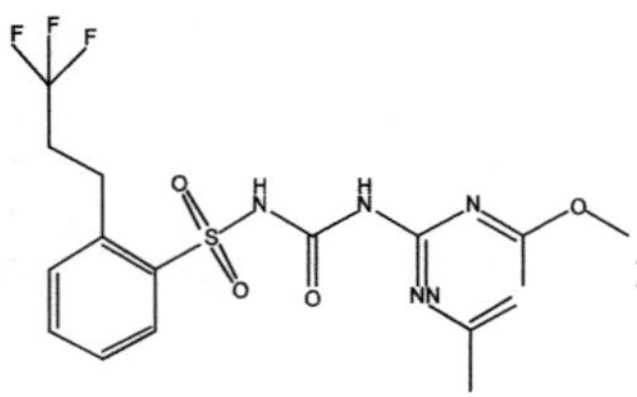	> 1000	Not classified	1.5	3.76	CGA 159902	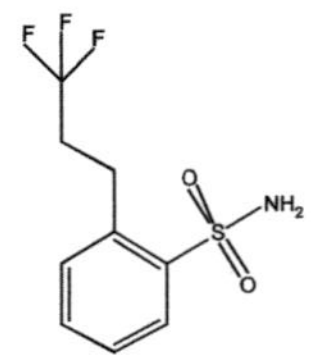	420	Harmful	1.81	10.02

Table 4 (continued)

Pesticide Name	Structure	Earthworm toxicity (mg/kg soil)	Hazard classification	Log K_{ow}^{a}	pK_a[a]	Transformation product Name	Structure	Earthworm toxicity (mg/kg soil)	Hazard classification	Log K_{ow}^{b}	pK_a[c]
Propoxycarbazone	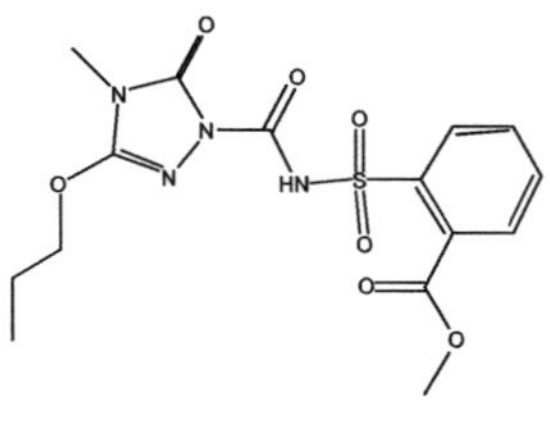	> 1000	Not classified	– 1.55	2.1	*N*-methylpropoxy triazolinone amide	No structure available	316	Harmful	–	–
Isoproturon	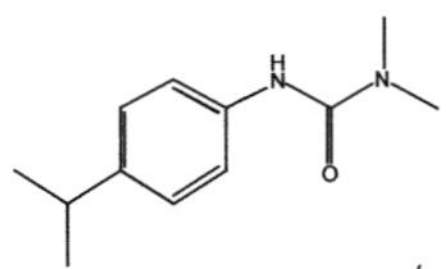	> 1000	Not classified	2.5	Not applicable	*N*-desmethyl-IPU	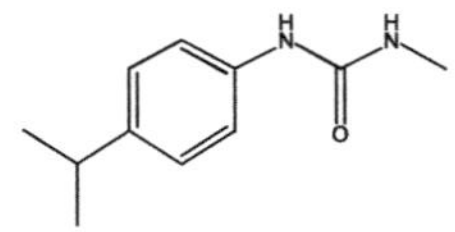	180	Toxic	2.54	0.51
Oxadiargyl	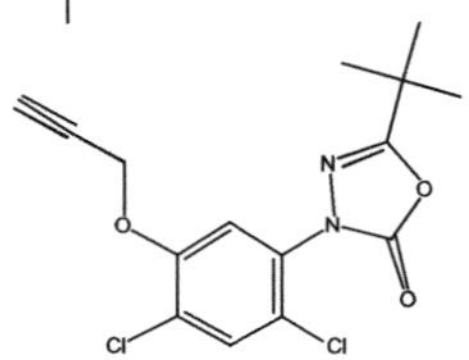	> 1000	Not classified	3.95	Not applicable	RP 025496	No structure available	120	Toxic	–	–
Methiocarb	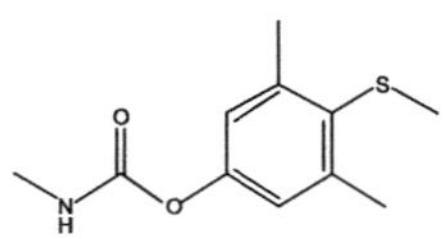	1322	Not classified	3.18	Not applicable	Methiocarb sulfoxide	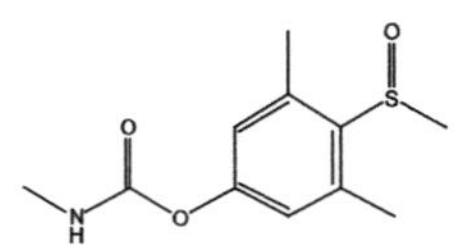	78	Toxic	0.88	10.39

Table 4 (continued)

Pesticide Name	Structure	Earthworm toxicity (mg/kg soil)	Hazard classification	Log K_{ow} [a]	pK_a [a]	Transformation product Name	Structure	Earthworm toxicity (mg/kg soil)	Hazard classification	Log K_{ow} [b]	pK_a [c]
Cyazofamid	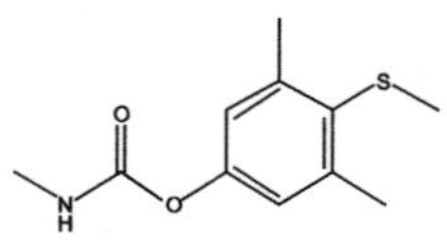	> 1000	Not classified	3.2	Not applicable	CCIM	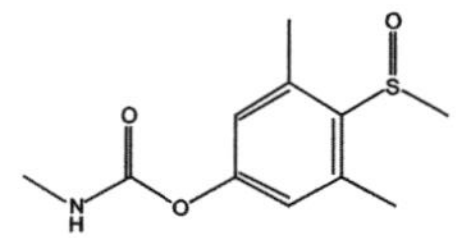	56	Toxic	3.04	7.18
Fenitrothion	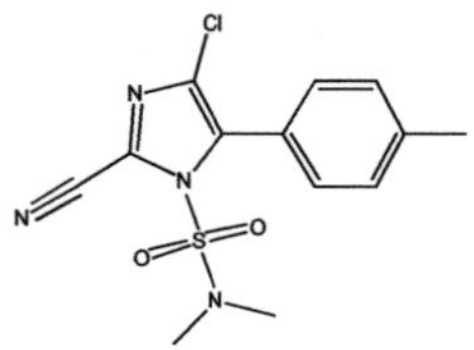	231	Toxic	3.32	Not applicable	3-methyl-4-nitrophenol	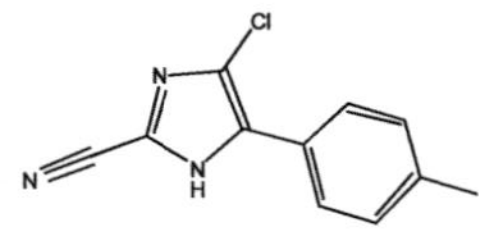	35	Very toxic	2.33	10.39
Chlorpyrifos	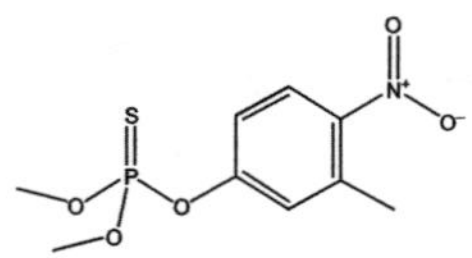	129	Toxic	4.7	Not applicable	3,5,6-trichloro-2-pyridinol	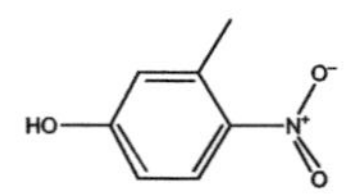	9.8	Very toxic	2.93	4.6

[a] Pesticide hydrophobicity and dissociation data from FOOTPRINT [24]
[b] Transformation product hydrophobicity calculated from the mean of *C* log *P* (Biobyte Corp.), KOWWIN (USEPA) and AlogPS (VCCLAB)
[c] Dissociation calculated using SPARC (University of Georgia)

only ten transformation products classified in a higher risk category than their parental pesticides (Table 4). Whilst an increase in hydrophobicity can explain some of the increases in potency to aquatic non-target organisms, this is not the case for changes in potency to earthworms. Generally transformation products that showed increase in earthworm toxicity compared to pesticides demonstrated reduced or equivalent hydrophobicity and increased dissociation based on available and estimated K_{ow} and pK_a data (Table 4). This is perhaps not surprising as an increase in hydrophobicity in a soil system will not only affect uptake into the worm but will also affect the degree of sorption to the soil matrix which will offset the i ncrease in bioconcentration factor from the pore water into the worm.

Two of the transformation products that demonstrate increased potency were the demethylated products of urea herbicides (3-(3-chloro-*p*-tolyl)-1-methylurea and *N*-desmethyl-isoproturon), whilst 3,5,6-trichloro-2-pyridinol and 3-methyl-4-nitrophenol are transformation products from OP insecticides that have had the ethylated and methylated phosphorothionate moiety cleaved from their structure, respectively.

3 Mechanisms of Increases of Toxicity for Transformation Products

From the previous sections and results of similar studies it is clear that when pesticides are considered, the majority of their transformation products elicit an equal or lower toxicity to aquatic non-target organisms and to earthworms than the parent pesticides [7, 25, 26]. A similar relationship has also been identified for biocides which is unsurprising as there is considerable overlap between chemicals considered as pesticides and those considered as biocides [27]. Generally it is unlikely that pesticidal transformation products will exhibit increased levels of toxicity to target organisms for which the parent pesticide has a specific mode of action towards that species. Propesticides can be expected to elicit this response but it would not be expected for transformation products considered a residue of the actual pesticide molecule. However, it is plausible, as has been demonstrated earlier that a few transformation products may show an increased potency to non-target organisms and this can be for organisms for which the parent compound may or may not elicit a specific mode of action. Increases in transformation product toxicity to aquatic non-target organisms when compared to parent pesticides can however often be attributed to increases in uptake, a change in mode of action, the presence of the parent toxicophore or the transformation product being the "active compound" of a pro-pesticides [7, 12]. These are discussed in more detail below.

3.1
Pro-pesticides

As discussed in the introduction, a number of pesticidal chemical groups are designed to act as pro-compounds. Generally this means that a precursor compound is applied during agricultural practice, which following uptake into the target organisms is ultimately metabolised to a more biologically active component that elicits the desired effect. Examples of pesticides and their associate active components are given in Fig. 2. These compounds are designed to act in this manner as the precursor molecule which can have favourable characteristics compared to the active molecule in terms of toxicity, selectivity, stability, biodegradability, mobility, persistence, sol-

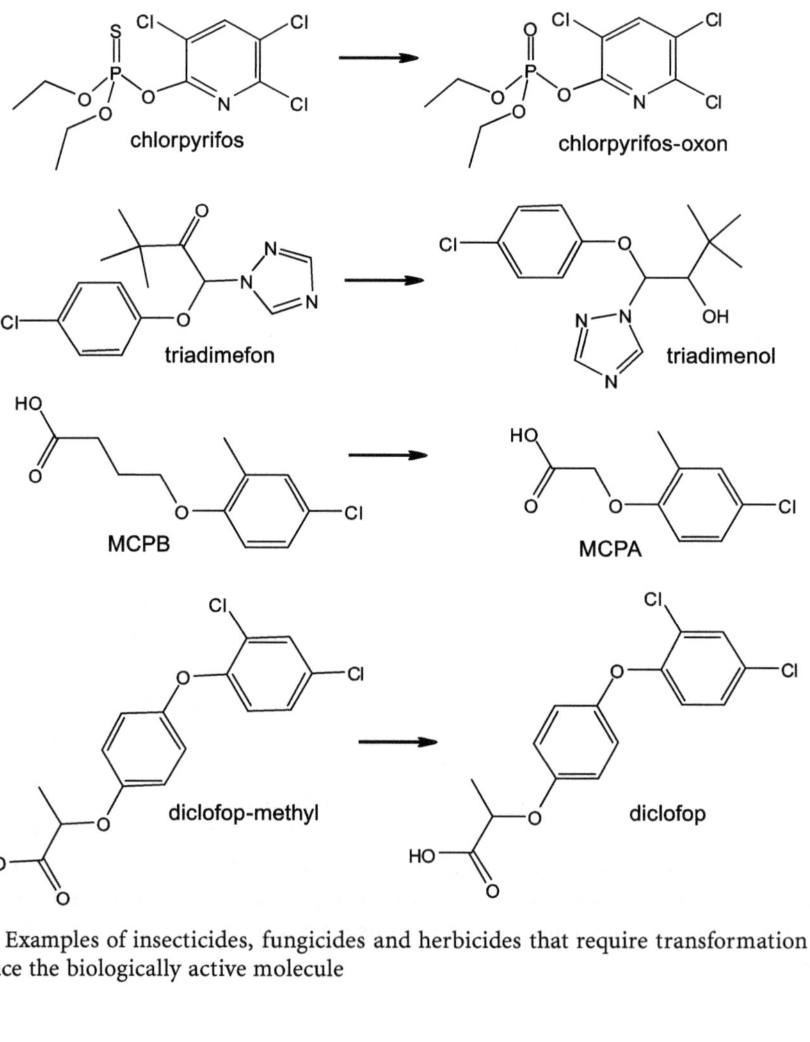

Fig. 2 Examples of insecticides, fungicides and herbicides that require transformation to produce the biologically active molecule

ubility and/or application. Insecticides from certain chemical classes work in this manner including carbamates, formamidines, thioureas, pyrethroids and organophosphates [13]. Diazinon a member of the thionophosate organophosphates undergoes oxidative desulphurisation as with other members of this chemical class and becomes a much more potent acetylcholinesterase inhibitor. The thionophate precursors are more stable, less volatile, have better insect cutical penetrability properties and lower mammalian toxicity than their biological active transformation products [2, 15].

Toxicological comparisons between the acute toxicity to non-target organisms of active molecule transformation products and their precursors can obviously be expected to demonstrate an increased potency of the active molecule to organisms that have the specific receptor site for the specific mode of action. For example, the active molecule of chlorpyrifos, chlorpyrifos-oxon, is at least an order of magnitude more potent, during acute LC_{50} studies, to the freshwater fish *Oryzias latipes* than the parent compound [28]. However, comparisons of this nature, as with some of the data points presented in Fig. 1 earlier, include organisms that contain the receptor site for the specific mode of action. It, therefore, cannot be considered a true demonstration of the extent of the difference in potency of precursors and active molecules because the toxicity exhibited by the precursor molecule toxicity will in fact be due to the active molecule following metabolic activation.

Pro-pesticide metabolic activation as well as occurring within a number of insecticidal classes is also observed in other pesticide classes. The systemic fungicide triadimefon, used for the control of powdery mildew and rusts in various fruits, vegetables and flowers, undergoes enzymic reduction in plants and fungi to the active molecule triadimenol which inhibits steroid demethylation [14, 15]. Unlike pro-insecticides which can have a specific mode of action which can affect aquatic invertebrates and fish the precursor and active molecule demonstrate similar levels of toxicity to daphnids and fish [29]. A range of herbicides are also applied as precursor molecules which require a transformation step before they become effective, these compounds are often applied in the form of esters, amides or salts which need to be hydrolysed to the free acid to become potent to target organisms. The aryloxyphenoxypropionates are one such group of herbicides which include diclofop-methyl. The arloxyalkanoic acid MCPA often applied as an herbicide itself is the active molecule of the inactive precursor MCPB which undergoes oxidative activation to MCPA which inhibits growth in broad-leaved weeds [5, 30].

3.2 Increases in Uptake

It is generally considered that the potency of organic chemicals to aquatic organisms can be attributed to one of four general chemical toxicity classes, with increasing potency: inert, less inert, reactive and specifically acting [31]. Inert chemicals exhibit a baseline toxicity known as non-polar narcosis, where the toxic effect is directly proportional and can be correlated to the partitioning behaviour of a chemical between the exposure media and the target organism. This non-specific reversible effect is independent of chemical structure and has been attributed to a range of chemicals in various taxa (e.g., [32, 33]). This toxic effect can be easily estimated and often hydrophobicity (i.e., $\log K_{OW}$) is used as an input parameter in developed quantitative structure–activity relationships (e.g., [34]). Therefore if we consider chemicals that act in a narcotic manner, if the chemical is transformed to a more hydrophobic molecule, it is likely that the transformation product will be more toxic than the parent compound. The nature of the test species will also be important, for example, an insecticide that acts via a specific mode of action to insects may exhibit narcotic action to green algae. Therefore increases in toxicity from a parent to a transformation product, neither of which exhibit a specific mode of action to the organism in question, can be due to a change in the uptake from parent to transformation product. A change in the extent of uptake will not only be influenced by the hydrophobicity of a chemical but also by its ability to cross cell membranes. The entry into cells for molecules that are highly dissociated is restricted when compared to similar un-dissociated molecules [35], therefore again when neither the pesticide nor the transformation product exhibit a specific action to a particular organism, the extent a compound is dissociated will influence the uptake and hence exhibited effect. Generally hydrophobicity and dissociation data are available for parent compounds as it is a requirement for registration, however these data are not always available for the transformation products but their are a range of quantitative structure property relationships available that can estimate hydrophobicity, e.g. ClogP (BioByte Corp.) and KOWWIN (US EPA/SRC) and dissociation, e.g. SPARC (University of Georgia). Some of these methods are discussed in the chapter by Howard.

3.3
Mode of Action

The toxicity of a specifically acting compound is due to an interaction between the molecule and a specific target site within an organism [36]. For example organophosphorus and carbamate insecticides inhibit the action of the enzyme acetylcholinesterase found in cholinergic nerve synapses by binding to the protein and thereby inhibiting the break down of the neurotransmitter acetylcholine [5]. A chemical would only be considered specifically acting if the organism under consideration contains the specific target site e.g., acetylcholinesterase. The majority of pesticides can be considered specifically acting when considering their desired action on target organisms but they can also have this action on non-target organisms. The two insecticide classes described above are effective insecticides but may pose a hazard to organisms with the site of action e.g., humans, fish and invertebrates.

The action of pesticides towards its intended target organism(s) (and some non-target organisms) would be considered as specifically acting because these organisms have a receptor that is effected by the chemical. However, if we consider organisms without that receptor site then the mode of action may be purely narcotic. For example, the urea herbicide diuron is used to control germinating grasses and broad-leaved weeds in a variety of crops and acts via inhibition of photosynthesis at the photosystem II receptor site [19]. Its intended target organisms possess this target site as do other non-target organisms such as green algae and aquatic macrophytes. Therefore diuron will exhibit increased potency to these organism because it will act on a specific site whilst organisms that do not contain this target site e.g., daphnids and fish may experience narcotic toxicity from this chemical. This is born out in the reported toxicity data for this chemical with acute toxicity to the green algal species *Scenedesmus subspicatus* reported at 0.019 mg/L (72 h E_rC_{50}) [37] whilst toxicity to the aquatic invertebrate *Daphnia magna* is approximately three orders of magnitude less at 12 mg/L (48 h LC_{50}) [19].

It is possible for transformation products to exhibit the same mode of action as the parent pesticide if during their breakdown, the structural moiety that exhibits the potency (toxicophore) within the pesticide is maintained within the structure of the transformation product. The majority of pesticides are members of a distinct pesticidal chemical class, pesticides within these classes generally have a common functional toxicophore associated with differing peripheral moieties e.g. Fig. 3. Some of the demethylated transformation products of urea herbicides exhibit similar levels of potency to non-target organisms that contain the target site of the parent pesticide. Desmethyl-chlorotoluron has a E_bC_{50} to an aquatic higher plant (*Lemna* sp.) within the same order of magnitude of the parent pesticide, 0.1 and 0.04 mg/L respectively [38]. Whilst the demethylated transformation product of diuron exhibits 75% of the herbicidal activity of diuron to the same taxa [39]. However, there are cases where even though a transformation product may contain the toxicophore of the parent pesticide it may still not exhibit the parental potency because the interaction between the toxicophore and target site may be inhibited/effected by the change in molecular shape or properties of the transformation product.

In some cases, the transformation reaction may give the transformation product a different and more potent mode of action than the parent compound. This is illustrated by the transformation reaction for carbaryl which is transformed to 5-hydroxy-1,4-naphthaquinone. Quinones are known to be highly toxic and act via enzymatically-based redox cycling resulting in superoxide generation and the reformation of the quinone [7]. The quinone functionality is not present in the parent compound.

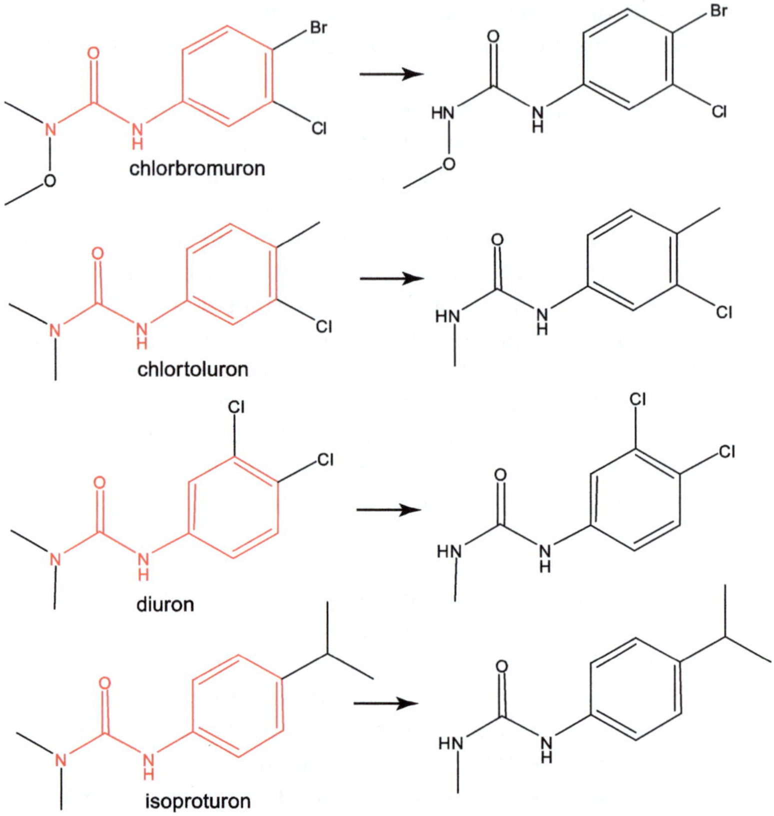

Fig. 3 Four urea herbicides with a sub-structural group common to all highlighted in red, all of which undergo a common demethylation degradation in soils

4
Comparison of Daphnid Ecotoxicity Estimation Techniques

It is clear from the previous sections that in some cases transformation products will be more toxic than their respective parent compound. The persistence and mobility of a transformation product may also be very different from its parent compound and, in some instances, will mean that the transformation product may persist in the environment for much longer than the parent compound. Compartments exposed to the transformation products may be different from the compartments exposed to the parent (see chapter by Hu et al., for a discussion of these issues). It is therefore critical that when assessing the risks of a chemical to the environment, we not only consider the parent compound, but we also consider the potential transformation products. As there are over 100 000 chemicals in use today, many of which will be transformed to a number of transformation products, it would be impossible to assess all of these transformation products experimentally. The development of predictive approaches, that can be used to identify those transformation products of most concern would therefore be highly beneficial. Therefore in the following section, we explore the predictive power of a range of predictive approaches for their use for assessing transformation products.

To undertake an assessment of the different approaches an experimental ecotoxicity data set was generated by randomly selecting fifty transformation products from the daphnid data set described previously. An assessment of the predictive performance of five techniques was undertaken by comparing predictions from each approach to the experimentally determined data. The data set included transformation product ecotoxicity data from a range of pesticide classes, i.e. insecticides, herbicides and fungicides; and chemical classes, e.g. organophosphorus insecticides, sulfonylureas and azoles.

4.1 Predictive Approaches

Five techniques were used to estimate daphnid acute ecotoxicity, these included freely available and commercial QSAR models, expert systems and an approach derived using data on structurally similar molecules called the "read-across approach".

ECOSAR is a freely available software system which matches the structure of a query molecule to one (or more) of its defined chemical class(es). For most classes, aquatic ecotoxicity values are then predicted using available linear correlations between toxicity and hydrophobicity, K_{OW} is estimated for the query molecule using KOWWIN (discussed in the chapter by Howard). The most recent version of ECOSAR (used in this study) contains over 150 relationships for approximately 50 chemical classes. For the purposes of assessing ECOSAR for predicting, transformation product toxicity, the structures of each of the chemicals in the data set were entered into the software system and in instances where the query compound was matched to one or more chemical classes, the most potent ecotoxicity estimate for daphnids was selected for comparative purposes.

TOPKAT (Accelrys Inc.) is a commercially available system and contains a range of cross-validated QSARs, which are multivariate statistical relationships between experimentally derived toxicity data and chemical descriptors that quantify chemical transport properties and biochemical interaction with the target site. It also provides the user with a measure of whether the query molecule fits within the prediction space of the chosen relationship and therefore whether the estimation is reliable. For this comparison exercise, estimated daphnid data were only compared if they fell within the optimum prediction space and all validation criteria were satisfied or if they fell outside the optimum prediction space but within a permissible range (as determined by TOPKAT). TOPKAT contains four separate models for the estimation of daphnid ecotoxicity.

In the read-across approach, chemicals that are structurally similar to the substance of interest are identified. Experimental data on the ecotoxicity of the structurally similar substances are then extracted from toxicological databases. This data is then used to give an indication of the potency of the substance of interest. The approach is, however, very reliant on a) similar compounds existing; and b) ecotoxicity data being available for these compounds. To assess the capability of read-across for estimating the toxicity of transformation products to daphnids, structurally similar compounds to the transformation products in the daphnid data set were identified using the similarity search function within the website ChemIDplus Advanced [40]. For the purposes of this study limits were set on the identification of similar chemicals; no chemicals were included if their similarity to the query compound was less than 70% (as defined by ChemIDplus) and only the 20 most similar chemicals were used. The similarity search function within ChemIDplus Advanced uses ISIS Direct Software (Elsevier MDL) for its chemical similarity searching. Once compounds were identified their CAS numbers were used to search for relevant ecotoxicological data within the ECOTOX database [18]. A read-across estimate of ecotoxicity for each transformation product was calculated by taking the mean of ecotoxicity values for all structurally similar compounds where a value was available.

The approach proposed by Escher et al. [41] was used to estimate the ecotoxic range of a transformation product, for the purposes of this evaluation the most potent extreme of that range was used as the prediction. This approach is discussed in more detail in the Chapter by Escher et al. (in this volume) but basically uses the principle of the toxic ratio [31] of the parent pesticide to estimate the toxic range and hence a maximum potency for a transformation product. The toxic ratio is the ratio between baseline toxicity, predicted using QSAR, and the toxicity determined experimentally for the end-point under investigation. Initially the baseline toxicity was estimated for each of the parent pesticides for the transformation products in the daphnid data set using a recommended non-polar narcotic QSAR [42]. This value was then used alongside the measured toxicity value for the parent compounds in order to derive the toxic ratio for each parent compound. The baseline toxicity of the transformation products was then estimated and this value was then used to estimate the specific (or most potent) ecotoxicity for a transformation product by multiplying it by the toxic ratio of the parent pesticide.

The final approach that was tested was one proposed by Sinclair and Boxall [7] which uses some identified differences between structural and physicochemical properties of the parent pesticide and its transformation product to allocate each transformation product an assessment factor. Parent pesticide ecotoxicity is then manipulated with the assessment factor to provide a conservative estimate of ecotoxicity for the transformation product. Assessment factors are allocated on the basis of parent pesticide potency, whether the transformation product has maintained the toxicophore of the parent pesticide and differences in uptake between parent and transformation product considering hydrophobicity and dissociation.

4.2 Evaluation of Predictive Ability

To evaluate the different predictive approaches, a refined version of the methodology used for comparing QSAR proposed by Moore et al. [43] was used. The ordinal ranking system was replaced by a ranking that provides a measure of the ability of a technique within each chosen statistic compared to the other techniques under evaluation. The approaches were ranked on their distance from the optimum summary statistic value standardized using the maximum distance from the optimum for all the techniques tested. An overall score was obtained by then calculating the mean of the individual rank scores, the best performing technique was identified as the one with a mean rank score nearest to zero, i.e. perfect performance. Genstat version 9.2 (VSN International Ltd.) and Excel 2000 version 9.0 (Microsoft) were used to analyse the data. The statistics generated for the predictive ability of each technique for the fifty transformation products are detailed below; those summary statistics used for the ranking system are identified with an asterisk (∗):

- actual number of compounds a technique could provide a prediction*
- percent positive deviation*
- mean absolute deviation*
- maximum absolute deviation
- minimum absolute deviation
- mean squared absolute deviation*
- percentage of compounds > 1 order of magnitude from experimental values*
- percentage of compounds > 2 orders of magnitude from experimental values
- percentage of compounds > 3 orders of magnitude from experimental values
- Pearson correlation coefficient*
- slope
- intercept.

Table 5 Summary statistics for the ability of five approaches to accurately estimate transformation product acute ecotoxicity to daphnids

Summary statistics	Optimum	ECOSAR	TOPKAT	Read-across	Sinclair and Boxall 2003 [7]	Escher et al. 2006 [41]
Number of compounds [a]	50	46	33	19	44	44
Summary statistic		1.3E – 01	5.5E – 01	1.0E + 00	1.9E – 01	1.9E – 01
% of positive deviations [a]	0	-10.9	19.7 [b]	13.2	11.4	2.3
Summary statistic		5.5E – 01	1.0E + 00	6.7E – 01	5.8E – 01	1.2E – 01
Mean absolute deviation (mol L^{-1}) [a]	0	140.2	0.6	6.6	0.9	0.8
Summary statistic		1.0E + 00	4.1E – 03	4.7E – 02	6.5E – 03	5.9E – 03
Maximum absolute deviation (mol L^{-1})	0	4096.0	4.4	54.5	8.3	14.7
Minimum absolute deviation (mol L^{-1})	0	3.3E-07	4.7E-05	7.9E-06	3.8E-06	6.4E-08
Mean square absolute deviation [a]	0	3.8E+05	1.3	3.2E+02	3.7	6.7
Summary statistic		1.0E + 00	3.5E – 06	8.5E – 04	9.9E – 06	1.8E – 05
% of compounds > 1 orders of magnitude [a]	0	56.5	48.5	63.2	70.5	52.3
Summary statistic		8.0E – 01	6.9E – 01	9.0E – 01	1.0E + 00	7.4E – 01
% of compounds > 2 orders of magnitude	0	41.3	21.2	52.6	27.3	20.5
% of compounds > 3 orders of magnitude	0	21.7	9.1	26.3	13.6	9.1
Pearson correlation coefficient [a]	1	0.417	0.085	0.450	-0.055	0.641
Summary statistic		5.5E – 01	8.7E – 01	5.2E – 01	1.0E + 00	3.4E – 01
Slope	1	0.001	0.170	0.004	-0.053	0.263
Intercept	0	0.4	0.5	0.1	0.6	0.3
Mean rank of summary statistics	0	0.67	0.52	0.52	0.46	0.23

[a] Mean rank derived from these statistics
[b] Positive deviation from 50% identified as significant (95% confidence limits)

The percentage positive deviation statistic is the percentage of the predictions that are over predicted from perfect correlation by a technique. If a predictive technique does not have a tendency to over or under predict values, i.e. over predicts as many values as it under predicts then you would expect the percentage positive deviation to be 50%. Therefore this statistic is used as a measure of the tendency of a package to over or under estimate potency. The data reported for this statistic is the distance from 50%, i.e. if positive the technique has a tendency to over-estimate the potency, if negative the technique has a tendency to under-estimate the potency whilst the further away from zero the more exaggerated this tendency. A one sample binomial test was used to identify if the identified tendency to under or over estimate the potency was significant at the 95% confidence limit.

Overall, based on the mean of the summary statistics, the approach proposed by Escher et al. [41] performed significantly better at predicting ecotoxicity for the whole daphnid data set with a score of 0.23 (Table 5). This approach was never ranked as the worst performing approach within any of the individual statistics for predictive ability. TOPKAT, the read-across approach and the approach of Sinclair and Boxall [7] demonstrated an equivalent overall ability with mean scores of 0.52, 0.52 and 0.46 respectively. Based on the overall ability, ECOSAR was the worst performing technique and was also the only technique to demonstrate an overall tendency to under-estimate the potency of transformation products to daphnids. The two expert systems and ECOSAR provided estimates for at least 88% of the compounds, whilst the requirement for estimates to fit within the optimum prediction space for TOPKAT limited the coverage of the data set to 66%. Due to the lack of ecotoxicity data for similarly structured compounds the read-across approach only managed to provide estimates for 19 transformation products (38%). The potency of carbendazim to daphnids, the biologically active component of the fungicide benomyl, was consistently under-estimated by all five techniques. With estimates being out by more than an order of magnitude. No transformation products were consistently over predicted by all five techniques.

Overall, the data indicate the method of Escher et al. [41] is the best performing approach. However, each of the methods adopts a different approach to assess toxicity so it may be appropriate to use a combination of methods when attempting to identify transformation products of potential concern.

5 Mixture Effects of Transformation Products

The data presented so far in the chapter has focused on single transformation products. However, within the natural environment an organism will not be exposed to a transformation product individually but will be exposed to a mixture of the parent compound and a range of its transformation products as well as other unrelated compounds (and their transformation products). Fenner et al. [44] proposed a simple approach to deal with this problem through the use of mixture risk quotients. The mixture risk quotient is based on the toxic unit approach and assumes that the parent compounds and their transformation products act in an additive manner. The mixture risk quotient is calculated using Eq. 1.

$$\mathrm{RQ} = \sum_{i=1}^{n} \frac{C_\mathrm{i}}{\mathrm{PNEC_i}} \tag{1}$$

where:

RQ = mixture risk quotient
C = concentration of the parent compound(s) or transformation product(s)
$\mathrm{PNEC_i}$ = Predicted no effect concentration for the parent compound(s) or the transformation product(s).

To explore the risk implications of mixtures of pesticides and their transformation products occurring in aquatic systems, we have taken data from a recent monitoring study of pesticides and transformation products in US streams [45]. This study looked at the occurrence of the parent compounds alachlor, metolachlor, acetochlor, dimethamid and atrazine as well as transformation products associated with each of these parent compounds. The monitoring was done in the spring and autumn at a number of sites.

In order to determine the mixture risk quotients for each of the study sites on each sampling occasion, data were obtained from the literature on the ecotoxicity of the parent compounds to fish and daphnids. Experimental ecotoxicity data were also obtained for the transformation products where available. In instances where experimental data were not available for the transformation products, estimates of ecotoxicity were obtained using the predictive approach of Escher which is described in the previous section. Experimental and predicted ecotoxicity data were then used alongside the monitoring data to calculate mixture risk quotients for each of the sampling sites on each sampling occasion for fish and daphnids. The mixture risk quotients were calculated a) using only the parent compound occurrence data; and b) the parent compound and transformation product data.

The results (Fig. 4) show that the risk quotients for the mixtures of parent compounds are all below unity, indicating an acceptable risk to fish and invertebrates at these study sites. Inclusion of the transformation products in the assessment increased the risk quotient very slightly and indicated that the transformation product and parent compound combination also posed an acceptable risk to fish and invertebrates. This is not sur-

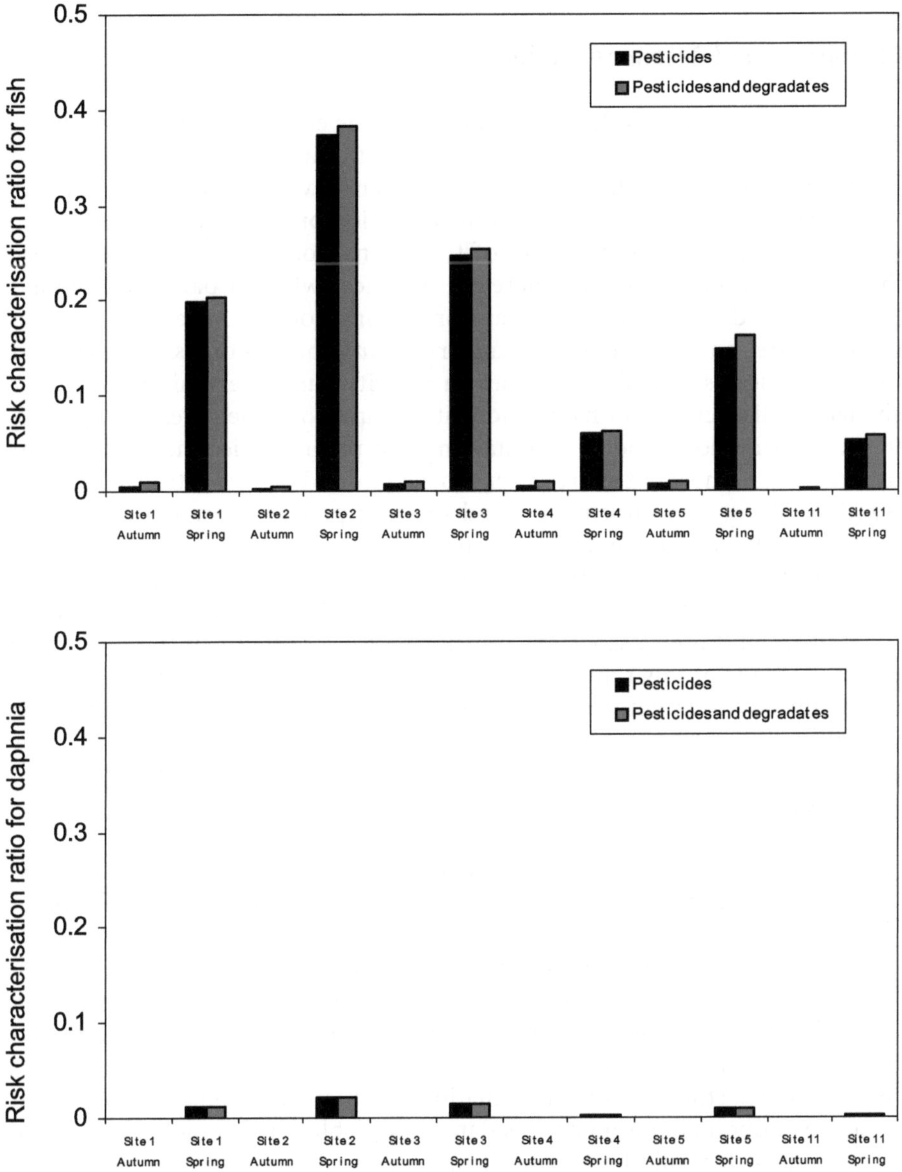

Fig. 4 Risk quotients for mixtures of pesticides and mixtures of pesticides and their associated transformation products occurring in stream in the US

prising in this particular case as the transformation products of the parent compounds are all less potent than the parent. Other pesticides, that transform to products that are more potent, could however give a different response.

6
Conclusions and Recommendations

While a large amount of data are available on the ecotoxicity of parent chemicals, much less information is available on the associated transformation products. Pesticides are the one class of substances where a large body of data are available on the ecotoxicity of transformation products. Using these data, it is clear that while the majority of transformation products are less toxic than their parent compound, there are instances where a parent compound is transformed to a more toxic transformation product. An examination of the structures and properties of transformation products that show increased toxicity indicates that the increases in toxicity can be explained by either changes in dissociation or hydrophobicity compared to the parent compound, the transformation product maintaining the parent toxicophore or the introduction of a new but more potent toxicophore during the transformation reaction. Using this information and existing predictive models, it is now possible to begin to predict the potential ecotoxicity of a transformation product based on its structure. In addition, using mixture toxicity models, it is possible to begin to estimate the potential combine effects of the parent compounds and their associated transformation products. In the future, the use of methods of this type, should enable the risks of transformation products to the environment to be better assessed. Compared to parent chemicals, we still however know very little about the environmental toxicity of these compounds so we would advocate that further work is done in the near future on the following aspects:

- Most work on transformation product ecotoxicity has been done on transformation products of pesticides. While some data are available for other transformation products (e.g. veterinary medicines, industrial chemicals and pharmaceuticals), these data are quite limited. We should begin to assess the effects of transformation products from these other groups and establish whether the relationships described in this chapter for pesticides hold true for the wider chemical universe.
- Ecotoxicity studies on transformation products have generally looked at acute endpoints and much less data is available on chronic and sublethal responses. It would be beneficial to generate data so that we can explore relationships between parent chronic toxicity and the chronic toxicity of the associated transformation products.
- Because of a scarcity of data, it is difficult to explore the mechanisms behind the increases in toxicity observed for earthworms. It is possible that the drivers for changes in toxicity in terrestrial systems are different from those in aquatic systems where the exposure route is less complex.
- Further work is required on the evaluation of the predictive models for estimating the ecotoxicity of transformation products. It may be appropri-

ate to combine aspects from a number of different approaches to estimate transformation product effects. The use of receptor-based models may also be valuable in providing information on whether a transformation product maintains that mechanism of action of the parent compound or not.

References

1. Walker CH, Hopkin SP, Sibly RM, Peakall DB (1997) Principles of Ecotoxicology. Taylor and Francis Ltd, London, UK
2. Coats JR (1993) Chemtech 23:25
3. Cremlyn RJ (1991) Agrochemicals: Preparation and Mode of Action. Biddles Ltd, Guildford, UK
4. Carson R (1963) Silent Spring. Hamish Hamilton, London, UK
5. Copping LG, Hewitt HG (1998) Chemistry and Mode of Action of Crop Protection Agents. The Royal Society of Chemistry, Cambridge, UK
6. Belfroid AC, Van Drunen M, Van Gestel CAM, Van Hattum B (1996) Relative Risks of Transformation Products of Pesticides for Aquatic Ecosystems. Institute for Environmental Studies and Institute for Inland Water Management and Waste Water Treatment, Amsterdam, Netherlands
7. Sinclair CJ, Boxall ABA (2003) Environ Sci Technol 37:4617
8. USEPA (2008) Pesticide Reregistration Status. http://cfpub.epa.gov/oppref/rereg/status.cfm?show=rereg, last accessed 17th January 2008
9. EU (2008) Technical Review Reports. http://ec.europa.eu/food/plant/protection/evaluation/new_subs_rep_en.htm, Accessed 17 Jan 2008
10. EU (2008) Technical Review Reports. http://ec.europa.eu/food/plant/protection/evaluation/existactive/list1-47_en.pdf, Accessed 17 Jan 2008
11. PSD (2008) ACP Published Evaluation Documents. http://www.pesticides.gov.uk/psd_evaluation_all.asp, Accessed 17 Jan 2008
12. Boxall ABA, Sinclair CJ, Fenner K, Kolpin D, Maund SJ (2004) Environ Sci Technol 38:368A
13. Drabek J, Neumann R (1985) Proinsecticides. In: Hutson DH, Roberts TR (eds) Insecticides. John Wiley & Sons, Chichester
14. Fedorov LA, Yablokow AV (2004) Pesticides: The Chemical Weapon That Kills Life. Pensoft, Moscow, Russia
15. Roberts T, Hutson D (1999) Metabolic Pathways of Agrochemicals, Part Two: Insecticides and Fungicides. The Royal Society of Chemistry, Cambridge
16. OECD (2004) Guidelines for the Testing of Chemicals, *Daphnia* sp. Acute Immobilisation Test, Organisation for Economic Co-Operation and Development. OECD, Paris, France
17. Maroni M, Auteri D, Grasso P, Alberio P, Redolfi E, Azimonti G, Giarei C, Visentin S (2002) Statistical Evaluation of Available Ecotoxicology Data on Plant Protection Products and Their Metabolites (Final Report), B1-3330/2001216. International Centre for Pesticides and Health Risk Prevention, Busto Garolfo, Italy
18. USEPA (2008) ECOTOX Database. http://www.epa.gov/ecotox/, Accessed 17 Jan 2008

19. Tomlin CDS (2000) The Pesticide Manual. BCPC, Farnham, UK
20. Tomlin CDS (2006) The e-Pesticide Manual. BCPC, Farnham, UK
21. PMRA (2008) Re-evaluation Documents. http://www.pmra-arla.gc.ca/english/pubs/reeval-e.html, Accessed 17 Jan 2008
22. EFSA (2008) Conclusions on the Peer Review of Pesticide Risk Assessments. http://www.efsa.europa.eu/en/about_efsa.html, Accessed 17 Jan 2008
23. Jensen J (1999) Terrestrial Hazard Classification of Toxic Substances: A Study to Evaluate Specific Terrestrial Hazard Criteria Using Pesticide and Biocide Toxicity Data, ECBI/19/99 Add. 7. National Environment Research Institute, Silkeborg, Denmark
24. FOOTPRINT (2008) The FOOTPRINT PESTICIDE PROPERTIES DATABASE, http://www.eu-footprint.org/ppdb.html, Accessed 17 Jan 2008
25. Belfroid AC, Van Drunen M, Beek MA, Schrap SM, Van Gestel CAM, Van Hattum B (1998) Sci Total Environ 222:167
26. Streloke M, Joermann G, Kula H, Spangenberg R (2002) Analysis of Toxicity Data on Aquatic Organisms for Regulatory Purposes. 2002. SETAC-Europe, 12–16 May, Vienna, Austria
27. Sinclair CJ, Boxall ABA (2002) Assessing the Environmental Properties and Effects of Biocide Transformation Products. Cranfield University, Bedfordshire, UK
28. Giesy JP, Solomon KR, Coats JR, Dixon KR, Giddings JM, Kenaga EE (1999) Rev Environ Contam Toxicol 160:1
29. EPA (2006) Reregistration Eligibility Decision for Triadimefon and Tolerance Reassessment for Triadimenol. Environmental Protection Agency, Washington, USA
30. Roberts T (1998) Metabolic Pathways of Agrochemicals, Part one: Herbicides and Plant Growth Regulators. The Royal Society of Chemistry, Cambridge
31. Verhaar HJM, Van Leeuwen CJ, Hermens JLM (1992) Chemosphere 25:471
32. Veith GD, Call DJ, Brooke LT (1983) Can J Fish Aquat Sci 40:743
33. Cleuvers M (2003) Toxicol Lett 142:185
34. Könemann H (1981) Toxicology 19:209
35. Esser HO, Moser P (1982) Ecotox Environ Safe 6:131
36. Escher BI, Hermens JLM (2002) Environ Sci Technol 36:4201
37. EFSA (2005) EFSA Sci Rep 25:1
38. EU (2005) Review Report for the Active Substance Chlorotoluron. European Commission, Brussels
39. Dewez D, Marchand M, Eullaffroy P, Popovic R (2002) Environ Toxicol 17:493
40. NATIONAL LIBRARY OF MEDICINE (2008) ChemIDplus. http://chem.sis.nlm.gov/chemidplus/, Accessed 17 Jan 2008
41. Escher BI, Bramaz N, Richter M, Lienert J (2006) Environ Sci Technol 40:7402
42. European Chemicals Bureau (2003) Technical Guidance Document on Risk Assessment in support of Comission Directive 93/67/EEC on Risk Assessment for new notified substances, Commission Regulation (EC) no. 1488/94 on Risk Assessment for existing substances, Directive 98/8/EC of the European Parliament and of the Council concerning the placing of biocidal products on the market, Part III. Brussels
43. Moore DRJ, Breton RL, MacDonald RB (2003) Environ Toxicol Chem 22:1799
44. Fenner KB (2001) Doctoral Thesis. Swiss Federal Institute of Technology, Zurich
45. Hladik ML, Roberts AL, Bouwer ES (2006) Chloroacetamide Herbicides and Their Transformation Products in Drinking Water. Report No. 91123. Awwa Research Foundation, Denver

Hdb Env Chem Vol. 2, Part (2009): 205–244
DOI 10.1007/698_2_015

Published online: 30 April 2008

Predicting the Ecotoxicological Effects of Transformation Products

Beate I. Escher[1] (✉) · Rebekka Baumgartner[1,2] · Judit Lienert[1] · Kathrin Fenner[1,2]

[1]Swiss Federal Institute of Aquatic Science and Technology (Eawag), PO Box 611, CH-8600 Dübendorf, Switzerland
escher@eawag.ch

[2]Institute for Biogeochemistry and Pollutant Dynamics, Swiss Federal Institute of Technology (ETH), ETH Zürich, CH-8092 Zürich, Switzerland

Abstract Persistent environmental transformation products are increasingly being detected in surface waters and previous parts of this volume have discussed methods for prediction and quantification. However, there is not sufficient experimental data on their ecotoxicological potential to assess the risk associated with transformation products, even if their occurrence and abundance is known. Herein, we review computational methods for the identification and prioritization of transformation products according to their ecotoxicological potential and specifically focus on the assessment of mixtures of organic environmental pollutants and their transformation products. These transformation products can be produced through abiotic or microbial degradation or from metabolism in higher organisms. The proposed model assumes concentration addition between the components of the mixture and uses Quantitative Structure Activity Relationships (QSARs) to fill data gaps. The model is illustrated for five pesticides and their environmental transformation products. Their overall toxic potential is derived by scaling predicted relative aquatic concentrations (RAC, see Fenner et al., 2008, in this volume) with the relative potencies of each transformation product followed by summing up the toxic potentials of all mixture components. The model is versatile and can also be used to assess the cocktail of metabolites that is excreted by humans and animals after consumption/ingestion of pharmaceuticals. The metabolites of pharmaceuticals and hormones that are excreted are often more hydrophilic and consequently presumably less toxic than the ingested parent compound. However, they may be more abundant and therefore may be relevant for overall risk assessment. The weak point of our method, as of any QSAR application, is the correct assignment of the mode of toxic action (moa) of transformation products because they do not necessarily exhibit the same moa as the parent compound. In the future, more emphasis must therefore be placed on this issue, e.g., by identifying toxicophores or other structural alerts that are indicative of a certain mode of toxic action. An improved mode of action assignment would make the model more robust. Nevertheless, the prediction method is valuable for screening purposes and for setting priorities for further experimental testing.

Keywords Baseline toxicity · Ecotoxicology · Environmental transformation products · Metabolites · Mode of toxic action · Pharmaceuticals · Pesticides · Herbicides · QSAR

Abbreviations

Baseline toxicity	Minimum toxicity of every chemical caused by nonspecific effects
EC_{50}	Effect concentrations leading to 50% of a specified maximum effect
f_{parent}	Fraction of parent after metabolism
f_i	Fraction of the given metabolite i
LC_{50}	Lethal concentration for 50% of the test species
moa	Mode of action or mode of toxic action
NOEC	No-observed effect concentration
RP_i	Relative potency of the given metabolite i, in relation to 100% potency of the parent compound
QSAR	Quantitative structure activity relationship
$TP_{mixture}$	Toxic potential of the mixture of parent compound and its metabolites
TP_{parent}	Toxic potential of the parent compound, which by definition equals 1
TR	Toxic ratio = ratio between the $EC_{50,\text{baseline}}$ and the experimental $EC_{50,\text{experimental}}$

1 Introduction

Many transformation products of environmental pollutants such as pesticides, biocides, pharmaceuticals, and industrial chemicals can be found in the environment [1]. In particular, transformation products of pesticides are often more abundant than their parent compounds ([1] and references cited therein). These observations triggered several studies on the risk assessment of transformation products. Some studies encompass in-depth investigations of degradation pathways, followed by synthesis of the identified transformation products and ecotoxicity testing [2–4]. This experimental/laboratory strategy was also recommended by Weyant and Pressel [5] in their evaluation of different strategies to deal with metabolites. However, given the overwhelming number of chemicals and the even higher number of possible transformation products, this approach is only feasible for selected case studies. Therefore, a lot of effort has been devoted to the development of simple screening methods for identifying transformation products of concern.

Belfroid et al. [6] proposed a scheme to identify whether transformation products of pesticides are likely to pose a lower, similar, or higher risk than their parent compound. Indicators used were presence and persistence in water and sediment and/or high toxicity/bioaccumulation potential. This approach allowed a relative risk ranking of metabolite mixtures, but only four out of 20 pesticides could be reliably ranked due to limited data availability.

Sinclair and Boxall [7] focused their screening method on identifying transformation products of pesticides that were more toxic than their parent. They concluded that the majority of transformation products are less toxic than their parent compound. Exceptions are products that are more hydrophobic and thus more bioaccumulative than their precursor, or those with a more potent mode of action. The latter can be explained as follows: (1) by the presence of a toxicophore that is formed during transformation of a propesticide into its active product, (2) the pesticide toxicophore remains intact during transformation but hydrophobicity increases, or (3) a different toxicophore is formed during the transformation. On the basis of these rules they developed a flow chart to select appropriate assessment factors that relate the toxicity of the parent compound to the predicted toxicity of the transformation product. This approach is valuable for preliminary hazard assessment and prioritization of further testing but cannot give a quantitative account of the risk associated with transformation products.

All of the above research deals with environmental transformation products of pesticides. However, there are also other groups of chemicals such as pharmaceuticals and hormones, biocides, consumer products, and industrial chemicals, which may also produce persistent transformation products. These products may be produced in the environment by abiotic or biotic processes or produced in humans or animals as a result of metabolism [5].

One can differentiate between three types of transformation products of environmental pollutants. First, environmental pollutants can be metabolized during the toxicokinetic phase of uptake/metabolism/distribution/elimination in organisms (Table 1). Here, the observed effect is actually due to the combined effect of different metabolites. Taking these transformation reactions into account will help to understand mechanisms of toxicity, species sensitivity differences, and time dependency of effects. Lee and Landrum [8, 9] developed a model to describe the mixture effects of PAH and their metabolites in *Hyalella azteca*. This combined toxicokinetic/toxicodynamic models convincingly demonstrated the importance of accounting for metabolite formation and how different mixture toxicity concepts can be incorporated into such models.

The second case refers to hormones, pharmaceuticals, and other compounds that are ingested, metabolized, and excreted by mammals (Table 1). Usually a hormone or pharmaceutical is extensively metabolized in the body and is excreted by mammals as a mixture of different metabolites. Although the general belief is that metabolism renders a drug more water soluble and consequently less hazardous for the aquatic environment, there are exceptions for pro-drugs and specifically acting metabolites. The third case refers to environmental transformation products of pesticides and other environmental pollutants (Table 1), which are formed both by abiotic and biotic transformation processes.

For assessing the risk from transformation products in the second and third case, one must, on the one hand, know which quantity of each of the different transformation products is present in the environment. On the other hand, one needs to know the toxic potential relative to the parent compound. Herein, we describe a simple prediction model for simulating the effects of mixtures of parent compounds and their transformation products. The model was developed for metabolites of human pharmaceuticals [11, 12] and will

Table 1 Cases in which transformation of environmental pollutants need to be accounted for

Case	Information source	Example
1. Metabolites formed in an organism during the toxicokinetic phase	Only implicitly, metabolites cannot be traced without mechanistic studies	Acetylcholine Esterase Inhibitors [10], PAHs [9]
2. Metabolites from mammalian metabolism excreted into wastewater and the environment	Pharmacokinetic information from the literature	Human pharmaceuticals [11, 12]
3. Environmental transformation products	Concentrations found in the environment	Pesticides [1], nonylphenol ethoxylates [13]

also be presented in those terms. Note, however, that the same approach can also be applied to environmental transformation products, as will be shown in the case studies for selected pesticides.

2 Model for Predicting the Ecotoxicological Effects of Transformation Products

2.1 General Outline of the Model

The model for predicting the ecotoxicological effects of mixtures of metabolites and their parent compounds assumes concentration addition of the effects of metabolites and their parent compound. If concentration addition holds and additionally all assumptions pertinent to the toxic equivalency concept apply [14], the toxic potential of the mixture of a parent compound and its metabolites, $TP_{mixture}$, is defined as the sum of the fraction of parent after metabolism, f_{parent}, and the product of the fraction of each metabolite i, f_i, scaled by the potency of the given metabolite RP_i, in relation to 100% potency of the parent compound (Eq. 1).

$$TP_{mixture} = f_{parent} + \sum_{i=1}^{n} f_i \cdot RP_i \,. \qquad (1)$$

The fractions of each metabolite f_i can be derived from measured environmental concentrations, from knowledge about the metabolites formed and excreted by an organism (e.g. often available in the pharmaceutical literature), or from predictions generated using fate and exposure models (see Fenner et al., 2008, in this volume). Examples are given in the case studies below. The computation of the relative potency of a metabolite, RP_i, is central to the model and is derived in detail in the next section.

The assumption of concentration addition is not a priori justified because only compounds with the same mode of toxic action will act concentration additive in mixtures [15]. If parent compound and metabolites act according to different modes of toxic action, the appropriate mixture toxicity model would be independent action [16]. A mixture of similarly and dissimilarly acting compounds would have to be assessed with a two-step model [17]. This approach is too complicated for our screening purposes. Moreover, the necessary information to distinguish between different modes of toxic action (moa) is missing in most cases. However, if there are only a small number of chemicals in a mixture, the predictions for concentration addition are often very similar to the observed effects in mixtures of compounds with different moa [18, 19]. Therefore, in our case of a mixture of one parent compound and a small number of metabolites, the use of concentration addition as so-called "realistic worst-case" scenario is justified.

2.2 Computation of the Relative Potency of the Metabolites RP_i

Experimental ecotoxicity data for metabolites are generally scarce. Therefore, the model relies on a large number of assumptions while using any experimental evidence available. If there are no toxicity data available at all, they are estimated by quantitative structure activity relationships (QSAR) as described below.

Ideally, toxicity data for the parent compound and all metabolites would be known. Typical toxicity data are effect concentrations leading to 50% of a specified maximum effect, EC_{50}. If the toxicity endpoint is lethality, the endpoint is termed LC_{50}, i.e. lethal concentration for 50% of the test species. Of course also data for chronic endpoints, e.g., no-observed effect concentrations for endpoints like reproduction, can be used. It is important, however, that data from the same biological species, the same incubation period, and the same endpoint are compared. This criterion limits the applicability typically to acute toxicity data because there is rarely a full set of chronic toxicity data available. However, in principle, the model is generally applicable to any endpoint be it chronic or acute, given that there are both a baseline QSAR and some experimental data for the parent compound available.

If experimental data for the parent compound and the metabolites are available, the relative potency of a given metabolite i can be calculated according to Eq. 2

$$RP_i = \frac{EC_{50,\text{parent}}}{EC_{50,i}} . \quad (2)$$

Unfortunately, in most cases these effect concentration data, $EC_{50,i}$, are not available for metabolites. The few cases, in which data for the metabolites are available, can be used for validation of the model. However, in general, the RP_i of metabolites has to be estimated based on QSARs. The concept is similar to the approached described in the work by Howard et al. (2008, in this volume) but a toxicity QSAR relates a physicochemical descriptor such as the octanol–water partition coefficient $\log K_{ow}$ or the liposome-water distribution ratio $\log D_{lipw}$ linearly to a toxicity endpoint, e.g. $\log EC_{50}$ [20–22]. Most available QSARs are limited to baseline toxicity. Baseline toxicity is the non-specific disturbance of the structure and functioning of the lipid bilayer of biological membranes [23]. Baseline toxicity, also called narcosis, constitutes the minimum toxicity of any compound in an organism. Compounds that exhibit considerably higher toxicity than this baseline toxicity are considered to be specifically acting. In our model we exclusively used baseline QSARs and rescaled them to specific toxicity where necessary as described below.

In the few cases where QSARs are available for specific modes of toxic action, they may be directly implemented in our model. QSARs for specific modes of action may include additional descriptors and multivariate statistics to account for differences in toxicodynamics.

Depending on the availability of experimental data, three cases for mixture toxicity prediction can be distinguished. Figure 1 depicts a flow diagram that can be used to select the appropriate case. The toxic ratio TR is used to determine whether a parent compound is a baseline toxicant or acts according to a specific mode of toxic action [24–26]. The TR is defined as the ratio between the $EC_{50,\text{baseline},\text{QSAR}}$ predicted by a QSAR (see next paragraph for the choice of appropriate QSAR equations) and the experimental $EC_{50,\text{experimental}}$ (Eq. 3). If the parent compound has a TR_{parent} of ≤ 10, it is a baseline toxicant, while a $TR_{\text{parent}} > 10$ points towards a specific mode of toxic action [25, 26].

$$TR_{\text{parent}} = \frac{EC_{50,\text{baseline},\text{QSAR},\text{parent}}}{EC_{50,\text{experimental},\text{parent}}} . \tag{3}$$

The next question in the flow chart of Fig. 1 deals with the toxicity of the metabolite. If no experimental data for the metabolite are available or the TR analysis for the metabolite (Eq. 4) and the parent compound reveal that they both are baseline toxicants, then case I applies.

$$TR_i = \frac{EC_{\text{baseline},\text{QSAR},i}}{EC_{\text{experimental},i}} . \tag{4}$$

Case I represents the simplest possible case where both parent and metabolite act merely as baseline toxicants. The derivation of $RP_{\text{baseline},i}$ for this case is given in Eq. 5 and visualized in Fig. 2a.

$$RP_{\text{baseline},i} = \frac{EC_{50,\text{baseline QSAR},\text{parent}}}{EC_{50,\text{baseline QSAR},i}} . \tag{5}$$

If the parent compound is specifically acting (TR > 10) and no toxicity data are available for the metabolite (case II), we assume that the true toxicity of the metabolite lies between exhibiting the same specific moa as the parent compound and baseline toxicity. In that case we compute a range of RP_i as depicted in Fig. 2b. Possible RP_i range between $RP_{\text{min},i}$ (Eq. 6) for the assumption of baseline toxicity and $RP_{\text{max},i}$ (Eq. 7) for the assumption that the metabolite exhibits the same moa as the parent compound. Note that Eq. 7 will result in the same value as Eq. 5.

$$RP_{\text{min},i} = \frac{EC_{50\ \text{specific},\text{parent}}}{EC_{50\ \text{baselineQSAR},i}} \tag{6}$$

$$RP_{\text{max},i} = \frac{EC_{50\ \text{specific},\text{parent}}}{EC_{50\ \text{specific},i}} \equiv \frac{EC_{50\ \text{baseline QSAR},\text{parent}}}{EC_{50\ \text{baseline QSAR},i}} . \tag{7}$$

In reality, the metabolite could even be more toxic than the parent compound, but due to a lack of experimental evidence in most cases we cannot account for this option quantitatively. However, Sinclair and Boxall [7] gave a number

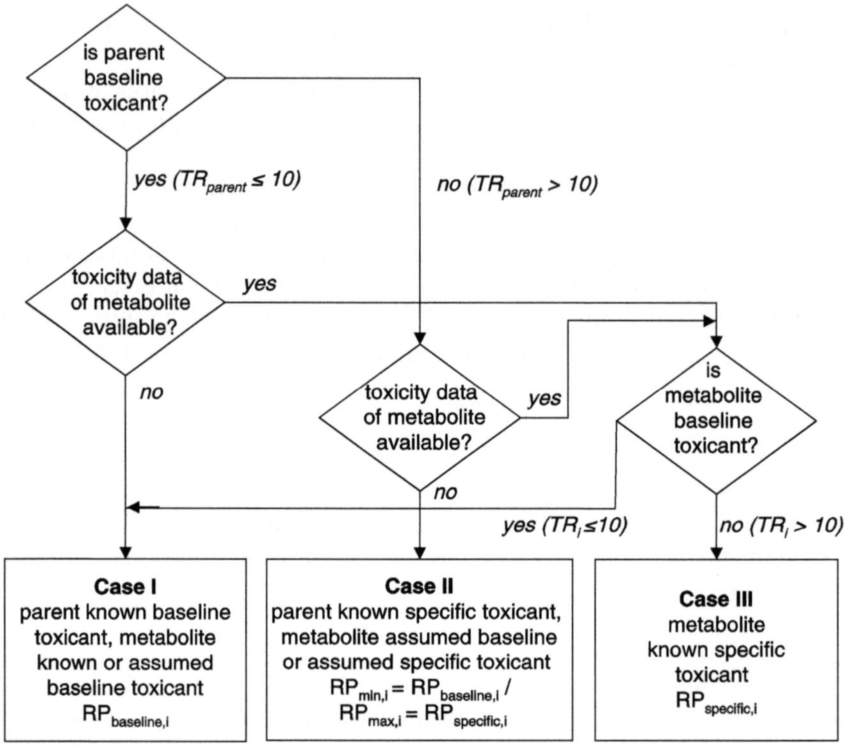

Fig. 1 Flow-chart for model selection and models to derive the relative potency of the metabolites RP_i in relation to 100% potency of the parent compound

of structural alerts that qualitatively indicate metabolites that might potentially be more toxic than their parent compounds. In our scheme, Case III covers a situation where experimental data for the metabolite is available and indicating a more potent moa than that of the parent compound.

Fig. 2 **a** Case I: Parent compound and metabolite are baseline toxicants. Note that, for consistency reasons, we used the computed $EC50_{baseline, parent}$ rather than the experimental value $EC50_{experimental, parent}$ to define $RP_{baseline, i}$. In this way, TR_{parent} is equal to 1 in all model calculations. The *empty circles* stand for estimated values, the *filled circles* for experimental data. **b** Case II: Parent molecule is specifically toxic and there is no information on the mode of toxic action of the metabolite. Therefore, for metabolite *i*, a minimum $RP_{min, i}$, representing baseline toxicity, as well as a maximum $RP_{max, i}$ were computed. $RP_{max, i}$ represents a metabolite that potentially exhibits the same specific mechanism as the parent compound. The *empty circles* stand for estimated values, the *filled circles* for experimental data. **c** Case III: The specific toxicity of the metabolits is confirmed by experimental evidence. The $RP_{specific, i}$ of the metabolite is calculated directly from the experimental data. The *empty circles* stand for estimated values, the *filled circles* for experimental data ▶

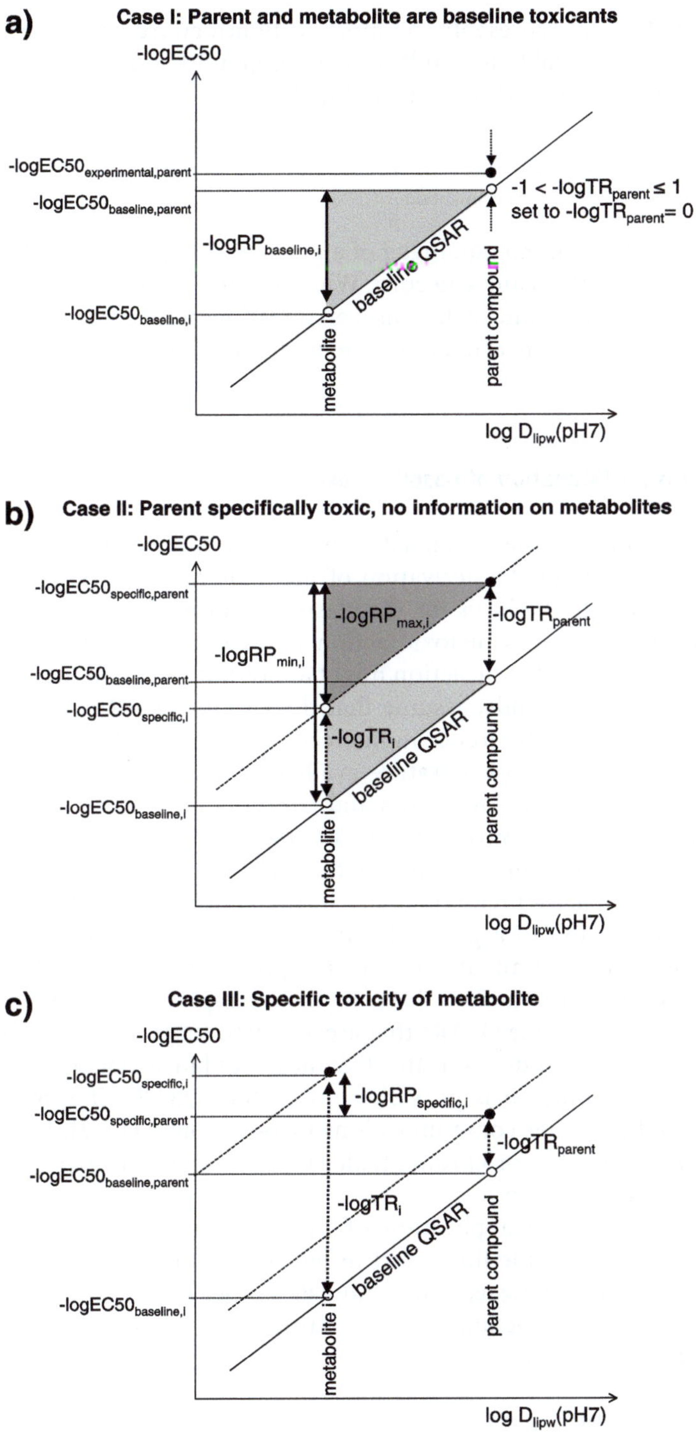
a)
Case I: Parent and metabolite are baseline toxicants
-logEC50
-logEC50$_{experimental,parent}$
-logEC50$_{baseline,parent}$
-logRP$_{baseline,i}$
-logEC50$_{baseline,i}$
-1 < -logTR$_{parent}$ ≤ 1
set to -logTR$_{parent}$= 0
baseline QSAR
metabolite i
parent compound
log D$_{lipw}$(pH7)
b)
Case II: Parent specifically toxic, no information on metabolites
-logEC50
-logEC50$_{specific,parent}$
-logRP$_{max,i}$
-logTR$_{parent}$
-logRP$_{min,i}$
-logEC50$_{baseline,parent}$
-logEC50$_{specific,i}$
-logTR$_i$
-logEC50$_{baseline,i}$
baseline QSAR
metabolite i
parent compound
log D$_{lipw}$(pH7)
c)
Case III: Specific toxicity of metabolite
-logEC50
-logEC50$_{specific,i}$
-logRP$_{specific,i}$
-logEC50$_{specific,parent}$
-logTR$_{parent}$
-logEC50$_{baseline,parent}$
-logTR$_i$
-logEC50$_{baseline,i}$
baseline QSAR
metabolite i
parent compound
log D$_{lipw}$(pH7)

Case III (Fig. 2c) represents the most straightforward case where experimental data are available for both parent compound and metabolite. In this case, RP_i can be derived directly from Eq. 7.

$$RP_{specific,i} = \frac{EC_{50\ experimental,\ parent}}{EC_{50\ experimental,i}} . \tag{8}$$

Note that the different metabolites i of a given parent compound do not necessarily all fall in the same category. While the toxic potential $TP_{mixture}$ can still be easily computed in such a case, error estimation becomes difficult due to varying degrees of uncertainty in the predictions for the different metabolites.

2.3 QSAR Models for Estimation of Baseline Toxicity

QSAR predictions of baseline toxicity are relevant in the context of our model because they allow for the derivation of the toxic ratio TR (Eqs. 3, 4). They therefore constitute the basis for the classification into nonspecific (= baseline) and specific modes of toxic action [25]. We do not further consider which specific mode of toxic action might apply, but rather, in the case of specifically acting compounds, assume that the specific QSAR is parallel to the baseline QSAR with a difference in intercept given by log TR (cf. Fig. 2a–c).

There is a vast body of literature available for the prediction of toxicity endpoints by QSARs (for overviews and concepts see [22, 27, 28]). Any toxicity QSAR is only valid for one specific biological species, exposure duration, toxicity endpoint, and mode of toxic action. The largest uncertainty in applying QSARs for toxicity predictions lies in the assignment of the appropriate mode of toxic action to a given chemical [29, 30]. Thus, the regulatory acceptance of QSARs is limited and current application in the EU is restricted to calculating baseline toxicity [31]. In the examples given here, we therefore only use the baseline QSARs that are accepted for regulatory applications in the EU and are laid down in the Technical Guidance Document (TGD) of the EU [31]. Of course it is possible to use other QSARs, if appropriate, as demonstrated in earlier communications by our group [11, 12]. Likewise, the application of a QSAR for chronic toxicity would allow the expansion of the model to chronic toxicity.

An overview of the application of the following Eqs. 8–12 for QSAR-estimation is given in the flow chart in Fig. 3. The three QSARs for baseline toxicity towards green algae, water, and fish that we used are listed in Table 2. They are based on the octanol–water partition coefficient $\log K_{ow}$ and are of the form given in Eq. 9:

$$\log EC_{50} = a \cdot \log K_{ow} + b . \tag{9}$$

Log K_{ow} is the most widely used descriptor for baseline QSARs but has many disadvantages in the context of our model. First, because octanol does not perfectly mimic the physicochemical properties of biological membranes, there are two different QSAR equations for nonpolar and polar compounds. This complication can be overcome by using the liposome–water partition coefficient log K_{lipw} as the descriptor instead [32]. We therefore recalculated the K_{ow}-based QSARs from the EU Technical Guidance Documents [31] using relationships between $\log K_{ow}$ and $\log K_{lipw}$ for nonpolar (Eq. 9) and polar compounds (Eq. 10), which had been experimentally determined by Vaes et al. [33, 34].

$$\text{nonpolar compounds: } \log K_{lipw} = 1.05 \cdot \log K_{ow} - 0.32\,, \tag{10}$$

$$\text{polar compounds: } \log K_{lipw} = 0.90 \cdot \log K_{ow} + 0.52\,. \tag{11}$$

Many metabolites but also some of the parent compounds are ionizable compounds. In this case it is not sufficient to assess the toxicity of the neutral species only, but the toxicity of the ionic species and the ratio of neutral and ionic species need to be additionally accounted for. It is widely accepted that the octanol–water distribution ratio of a charged organic chemical is not a good estimate of its bioaccumulation potential because the anisotropic features of biological membranes accommodate charged species better than octanol does [35]. Again, liposome–water distribution ratios are a better descriptor for toxicity calculations. Therefore, we replaced $\log K_{lipw}$ with the liposome–water distribution ratio at pH 7, $\log D_{lipw}$ (pH7) (Eq. 11). The D_{lipw} (pH7) is the sum of the products of the fraction of a given species j and the K_{lipw} of this species (Eq. 11).

$$D_{lipw}(\text{pH7}) = \sum_j f_j \cdot K_{lipw,j}\,. \tag{12}$$

Often the K_{lipw} of charged species is not available. In a literature review, we summarized evidence that the K_{lipw} of a charged molecule is typically about one order of magnitude lower than that of the corresponding neutral species [35]. This leads to Eq. 12 for the estimation of D_{lipw} of ionizing species in those cases where the K_{lipw} of the charged species is not known.

$$D_{lipw}(\text{pH7}) = f_{neutral} \cdot K_{lipw,\,neutral} + \sum_j f_{j,charged} \cdot \frac{K_{lipw,neutral}}{10}\,. \tag{13}$$

The accordingly rescaled QSARs are listed in Table 2. They were derived from the K_{ow}-based QSAR for nonpolar narcosis because this QSAR is based on a larger data set than the QSAR for polar narcosis. However, if all assumptions are correct, the resulting D_{lipw}-based QSARs should not be different from the one derived from the Kow-based QSAR for polar narcotics. This is indeed the case for fish but not for *Daphnia magna*. In fish, the slope is equal for both models and the intercept of the rescaled QSAR derived from the K_{ow}-based

Table 2 Baseline toxicity QSARs

Biological species	Toxicity endpoint	QSAR from TGD[a]	Rescaled QSAR[b]
Green algae *Pseudokirchneriella subcapitata*[c]	72–96 h EC_{50} growth inhibition	$\log(1/EC_{50}(M)) = 1.00 \cdot \log K_{ow} + 1.23$	$\log(1/EC_{50}(M)) = 0.95 \cdot \log D_{lipw} + 1.53$
Water flea *Daphnia magna*	48 h EC_{50} immobilization	$\log(1/EC_{50}(M)) = 0.95 \cdot \log K_{ow} + 1.32$	$\log(1/EC_{50}(M)) = 0.90 \cdot \log D_{lipw} + 1.61$
Fish *Pimephales promelas*	96 h LC_{50} lethality	$\log(1/LC_{50}(M)) = 0.85 \cdot \log K_{ow} + 1.39$	$\log(1/LC_{50}(M)) = 0.81 \cdot \log D_{lipw} + 1.65$

[a] QSAR from [31], citing [36, 37]
[b] rescaled with Eqs. 9, 10
[c] formerly known as *Selenastrum capricornutum*

QSAR for polar narcotics differ by only 0.09 log units from the one presented in Table 2. In contrast, in the case of *Daphnia magna*, the slope is by a third lower and the intercept by a third higher when the rescaled QSAR is based on the K_{ow}-based QSAR for polar narcotics. For the green algae *Pseudokirchneriella subcapitata* only the K_{ow}-based QSAR for nonpolar narcotics is listed in the TGD [31]; therefore such a comparison is not possible.

The $\log D_{lipw}$-based QSARs in Table 2 are universally applicable for polar, nonpolar, and charged compounds but in the derivation of the appropriate physicochemical descriptors the information on polarity and speciation must be included in a way that is shown below.

3 Derivation of the Physicochemical Properties Used in the QSAR Model

Ideally, experimental log D_{lipw} (pH7) values should be used for EC_{50} calculations (see Fig. 3). Unfortunately, experimental data are available for a limited number of compounds only, mostly pharmaceuticals. If the $\log K_{lipw,j}$ values of the different species are available, the $\log D_{lipw}$ (pH7) can be calculated according to Eq. 12, provided that the acidity constant(s) are available to derive the fractions of the different chemical species. If only $\log K_{lipw}$ of the neutral species is available, one can use Eq. 13 for the calculation of $\log D_{lipw}$. The flow chart in Fig. 3 shows how to proceed for a given data availability situation. If several types of input data are available, it is recommended to use the experimental data with highest quality. If several experimental data sets of apparently equal quality are available, we suggest using the geometric mean of the data. If no experimental data are available at all, the K_{ow} and the pK_a have to be estimated with prediction models. For estimating pK_a SPARC [38]

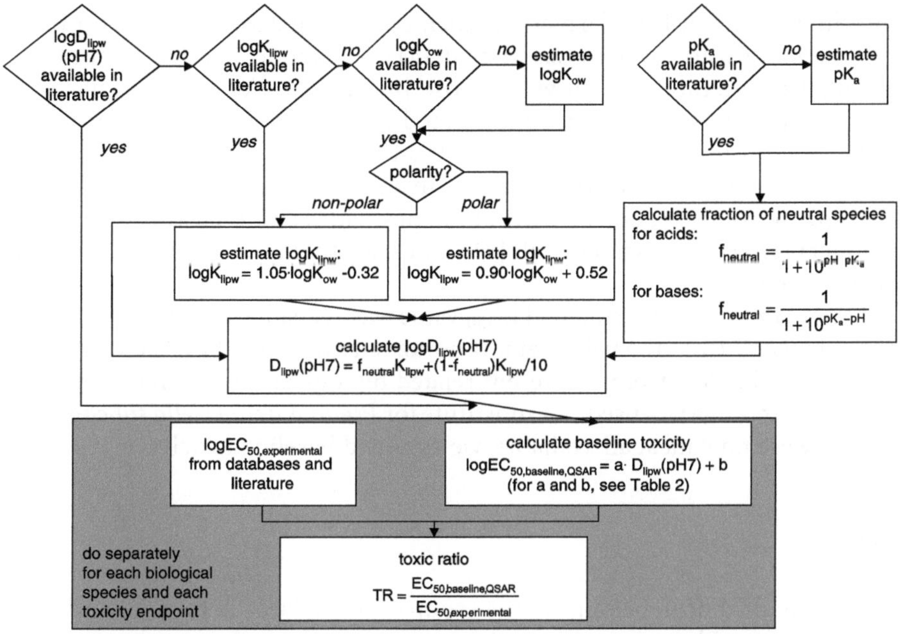

Fig. 3 Flow chart for selection of physicochemical parameters and their conversion into the input parameter for the QSARs, log D_{lipw} (pH7)

is widely used, whereas KOWWIN v1.67 from EPISUITE [39] is often used for the estimation of K_{ow}. For very complex molecules, where metabolism only changes a small part of the molecule, it is often preferable to estimate the K_{ow} in an incremental fashion based on the known K_{ow} of the precursor. The fragment method developed by Hansch et al. [22, 40] can be used for this purpose and is also implemented in KOWWIN v1.67 from EPISUITE [39].

4 Case Studies

4.1 Mixtures of Pesticides and Their Environmental Transformation Products

4.1.1 General Remarks

Knowledge on transformation products and their compound properties is well developed for pesticides. In the following we therefore illustrate our approach for the following five pesticides (four herbicides and one fungicide) and their transformation products: diuron, atrazine, dicamba, bromoxynil, and chlorothalonil. Transformation schemes and exposure estimates for these five case studies are given in Gasser et al. [41]. The model to calculate the toxic potential of mixtures, which we propose in Eq. 1, works in a modular way, i.e. any exposure estimate can be combined with its effect estimates, provided that the exposure estimates yield the fractions of transformation products and parent compound in the mixture. In our case study, we used the relative aquatic concentrations (RAC) of the pesticides and their transformation products reported in Gasser et al. [41] and scale them to a total of 1 to obtain the fractions f_i.

If more than one experimental EC_{50} value was available, the geometric mean was calculated from the valid data. If no toxicity data for the selected species were available, data from a different related biological species were used, e.g. another green alga was used as a substitute for *Pseudokirchneriella subcapitata*. If there were no data at all available, we assumed baseline toxicity.

4.1.2
Diuron

The herbicide diuron belongs to the group of phenylurea herbicides and is an inhibitor of the Photosystem II of algae and other plants. It is used not only as an agricultural herbicide but also as a biocide in paints and other consumer products. Diuron is a neutral compound and all its transformation products are neutral too (Fig. 4). All transformation products exhibit hydrophobicities within the same order of magnitude as the parent compound. DCPMU and DCPU are the products of demethylation on the amine group and MCPDMU has lost one chlorine substituent. All experimental physicochemical data were obtained from [39] and are given in Fig. 4.

Tixier et al. [4] have identified, synthesized, and assessed the toxicity of all transformation products of diuron. The bioassay they used was a bioluminescence inhibition test with the marine bacterium *Vibrio fischeri*. Since diuron does not exhibit any specific mode of toxic action towards bacteria, the QSAR analysis using a rescaled QSAR for *Vibrio fischeri* [42] only confirmed that diuron and all its metabolites with the exception of DCA (3,4-dichloroaniline) act as baseline toxicants (Table 3, for full names of metabolites see Fig. 4). However, DCA was 46-times more toxic than predicted with the baseline toxicity QSAR and almost two orders of magnitude more toxic than the parent compound. Such a specific mode of toxic action of a transformation product cannot easily be predicted unless toxicophores like the aniline structure present in DCA are considered as a signal. This is discussed in the conclusion section in more detail.

Fig. 4 Degradation pathway, physicochemical descriptors, and toxicity model of diuron. ▶ Data are from (*a*) [43], (*b*) [39], (*c*) [44], (*d*) [45]

Diuron

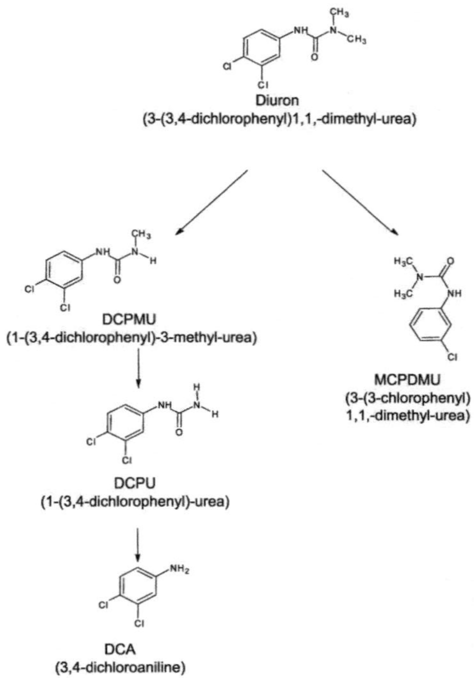

Physicochemical Parameters	pK_a	$\log K_{ow}$		$\log D_{lipw}$(pH 7)
Diuron	-	2.68	(a)	2.93
DCPMU	-	2.94	(b)	3.17
MCPDMU	-	2.00	(b)	2.32
DCPU	-	2.65	(b)	2.91
DCA	-	2.69	(b)	2.94

Green algae (*Pseudokirchneriella subcapitata*)	$\log(1/EC_{50}(M))$ baseline	$\log(1/EC_{50}(M))$ experimental		TR		$RP_{i,specific}$ or $RP_{i,max}$	$RP_{i,baseline}$ or $RP_{i,min}$
Diuron	4.32	6.71	(a)	248	specific moa	1.00	
DCPMU	4.54	6.58	(d)	110	specific moa	0.74	
MCPDMU	3.73	n.a.		n.a.		0.26	0.001
DCPU	4.29	n.a.		n.a.		0.94	0.004
DCA	4.32	4.51	(c)	2	baseline		0.004

Water flea (*Daphnia magna*)	$\log(1/EC_{50}(M))$ baseline	$\log(1/EC_{50}(M))$ experimental		TR		$RP_{i,specific}$ or $RP_{i,max}$	$RP_{i,baseline}$ or $RP_{i,min}$
Diuron	4.25	4.70	(a,c,d)	3	baseline		1.00
DCPMU	4.46	6.74	(d)	191	specific moa	110	
MCPDMU	3.70	n.a.		n.a.			0.28
DCPU	4.22	n.a.		n.a.			0.95
DCA	4.26	5.55	(c)	20	specific moa	7.08	

Fish (*Pimephales promelas*)	$\log(1/EC_{50}(M))$ baseline	$\log(1/EC_{50}(M))$ experimental		TR		$RP_{i,specific}$ or $RP_{i,max}$	$RP_{i,baseline}$ or $RP_{i,min}$
Diuron	4.02	4.22	(b)	1.6	baseline		1.00
DCPMU	4.21	n.a.				1.55	0.99
MCPDMU	3.53	n.a.				0.32	0.20
DCPU	4.00	n.a.				0.95	0.61
DCA	4.03	4.32	(b)	1.9	baseline		1.02

Table 3 Toxicity of diuron and its environmental transformation products in the bioluminescence inhibition test with *Vibrio fischeri*. Baseline toxicity estimates stem from physicochemical data presented in Fig. 4, baseline QSAR are given in [42], and experimental toxicity data in [4]; moa = mode of action

	$\log(1/EC_{50}(M))$ baseline	$\log(1/EC_{50}(M))$ experimental	Toxic ratio TR	moa classification	Relative potency RP_i
Diuron	3.86	3.54	0.48	baseline	1.00
DCPMU	4.04	4.08	1.08	baseline	3.45
MCPDMU	3.37	3.48	1.28	baseline	0.88
DCPU	3.83	4.14	2.00	baseline	3.97
DCA	3.86	5.53	46	specific moa	97

For all three toxicity endpoints selected in Table 2, a lot of experimental toxicity data are available for diuron and its transformation products (Fig. 4). Such a sound experimental database is rare, even in the case of pesticides, but certainly in the case of most pharmaceuticals.

As expected, diuron has a high specific toxicity towards algae with a TR of 250. Interestingly, also the first transformation product, DCPMU, still has a specific effect towards algae with a TR of 110. In water flea, the parent compound is a baseline toxicant while DCPMU has an even higher specific effect than in algae. Consequently, for algae the $RP_{i,\text{specific}}$ of the metabolite DCPMU is 0.74 but for water flea it is 110. This is an unexpected result and should be confirmed with more experimental data.

Since we do not have any toxicity data on MCPDMU and DCPU, we can only give ranges of RP_i, which are rather wide because of the high specific toxicity of the parent compound. If one would want to prioritize further studies there is clearly a need for more experimental data for these two transformation products.

DCA is a baseline toxicant in algae and fish but specifically toxic in *Daphnia magna*. Urrestarazu-Ramos et al. [46] compared the sensitivity towards different aromatic amines between different aquatic species and concluded that water fleas are consistently more sensitive than other aquatic species. However, they could not resolve the underlying mechanism.

For calculating the toxic potential of each metabolite TP_i and of the mixture of parent compound and metabolites, TP_{mixture} (Eq. 1), the relative fractions f_i of each transformation product need to be known. As described earlier we used the relative aquatic concentrations (RAC) calculated by Gasser et al. [41]. These data were used here after rescaling to fractions. The resulting f_i are depicted in Fig. 5, left column. Approx. 50% of diuron is still present as parent compound after metabolism while DCPMU and MCPDMU are quantitatively the most relevant transformation products, both contributing approx. 20% of f_i to the total mixture. The RP_{parent} by

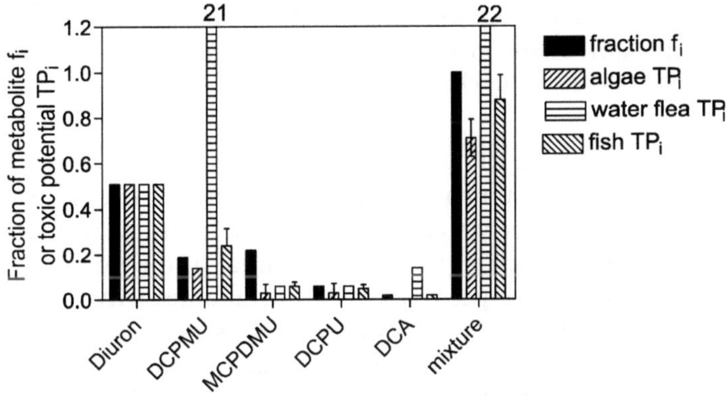

Fig. 5 Fractions of metabolites formed, f_i, and toxic potential of the metabolites, TP_i, and the mixture, $TP_{mixture}$, for diuron. *Bars* correspond to the average between $TP_{i,min}$ and $TP_{i,max}$ and the *error bars* indicate the range between minimal and maximal estimated toxic potential

definition equals one and thus TP_{parent} is equal to f_{parent}. When the fractions of the metabolites, f_i, are scaled with the relative potencies, RP_i, to derive the toxic potentials of the metabolites, TP_i, it becomes evident how much the RP_i of each compound contributes to the overall effect of the mixture: While f_i of DCPMU and MCPDMU are very similar, their TP_i and thus their contribution to the overall effect are very different. DCPMU is more toxic in all bioassays than MCPDMU but the difference is most pronounced in *Daphnia magna* with a TP_{DPCMU} of 21 (Fig. 5). Consequently, DCPMU dominates the toxic potential in water flea while the $TP_{mixture}$ towards the other species is the result of a more equal contribution from all metabolites (Fig. 5).

4.1.3
Atrazine

Atrazine is also an inhibitor of Photosystem II [47]. It is still used as herbicide in corn (*Zea mays*) in Switzerland and the U.S., but has been banned in the European Union. The toxicity of the mixture of atrazine and its transformation products is easier to understand than that of diuron. Atrazine and the transformation products from dealkylation of the amine group, DEA and DIA (Fig. 6), are specifically toxic towards algae but baseline toxic for water flea and fish. Accordingly, RP_i values can be calculated for DEA and DIA. In HA the chlorine substituent is replaced by a hydroxy-group. The RP_i of HA in algae can only be estimated to range between 0.004 and 0.36, depending on its mode of action. There is a possibility that the transformation products also act specifically in water flea and fish, but, again, information to corroborate this hypothesis is missing.

Fig. 6 Degradation pathway, physicochemical descriptors, and toxicity model of atrazine. ▶ Physicochemical properties are from (*a*) experimental value from Episuite [39], (*b*) estimated values from Episuite [39], (*c*) [44], (*d*) [48]

The combination of the RP_i with f_i derived from the RAC given in Gasser et al. [41] yield the toxic potentials depicted in Fig. 7. The transformation product DEA is quantitatively significant, but, due to its low relative potency, it does not contribute much to the toxic potential of the mixture. The contribution of the quantitatively less prevalent metabolites DIA and HA is even lower. Consequently, the $TP_{mixture}$ is dominated by the parent compound. It can thus be concluded that, in the case of atrazine, the contribution of the transformation products to the total aquatic risk is negligible.

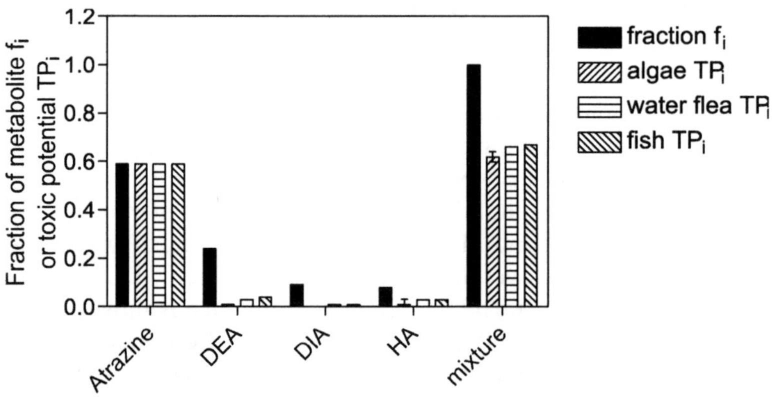

Fig. 7 Fractions of metabolites formed, f_i, and toxic potential of the metabolites, TP_i, and the mixture, $TP_{mixture}$, for atrazine. *Bars* correspond to the average between $TP_{i,min}$ and $TP_{i,max}$ and the *error bars* indicate the range between min and max. If there is no *error bar*, the mode of action is either known or baseline toxicity applies

4.1.4
Dicamba

The phenoxyacetic acid dicamba is used as a herbicide. It acts through inhibition of the synthesis of the phytohormone auxine, a specific mechanism related to plant growth. Accordingly, it is highly specifically toxic towards algae with a TR of 6500 but a baseline toxicant in water flea (Fig. 8). No toxicity data were available for fish. None of its diprotic and triprotic transformation products have been investigated for their toxicity. Consequently, their modeled ranges of RP_i cover more than three orders of magnitude, pointing towards a high priority for future testing. Also there is considerable uncertainty related to the toxicity estimate of 3,6-dichlorosalicylic acid because this compound is predominantly present in the double negatively charged form.

Atrazine

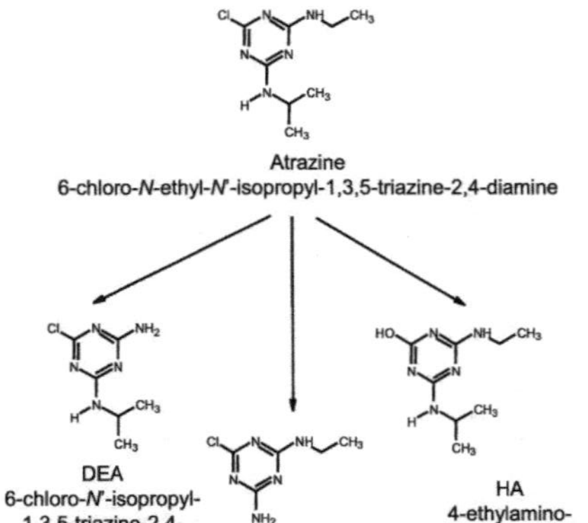

Atrazine
6-chloro-*N*-ethyl-*N'*-isopropyl-1,3,5-triazine-2,4-diamine

DEA
6-chloro-*N'*-isopropyl-1,3,5-triazine-2,4-diamine

DIA
6-chloro-*N*-ethyl-1,3,5-triazine-2,4-diamine

HA
4-ethylamino-6-isopropylamino-1,3,5-triazine-2-ol

Physicochemical Parameters				
	pK_a	$logK_{ow}$		log D_{lipw}(pH 7)
Atrazine	1.71	2.61	(a)	2.87
DEA	1.65	1.51	(a)	1.88
DIA	1.58	1.15	(a)	1.56
HA	5.15	2.09	(b)	2.40

Green algae (*Pseudokirchneriella subcapitata*)	log(1/EC_{50}(M)) baseline	log(1/EC_{50}(M)) experimental		TR		$RP_{i,specific}$ or $RP_{i,max}$	$RP_{i,baseline}$ or $RP_{i,min}$
Atrazine	4.26	6.20	(c)	88	specific moa	1.00	
DEA	3.32	4.80		31	specific moa	0.04	
DIA	3.01	4.57		37	specific moa	0.02	
HA	3.81	n.a.				0.00	0.36

Water flea (*Daphnia magna*)	log(1/EC_{50}(M)) baseline	log(1/EC_{50}(M)) experimental		TR		$RP_{i,specific}$ or $RP_{i,max}$	$RP_{i,baseline}$ or $RP_{i,min}$
Atrazine	4.19	3.93	(c)	0.5	baseline		1.00
DEA	3.30	n.a.					0.13
DIA	3.01	n.a.					0.07
HA	3.77	n.a.					0.38

Fish (*Pimephales promelas*)	log(1/EC_{50}(M)) baseline	log(1/EC_{50}(M)) experimental		TR		$RP_{i,specific}$ or $RP_{i,max}$	$RP_{i,baseline}$ or $RP_{i,min}$
Atrazine	3.97	4.16	(c)	1.5	baseline		1.00
DEA	3.17	n.a.					0.16
DIA	2.91	n.a.					0.09
HA	3.59	n.a.					0.42

Fig. 8 Degradation pathway, physicochemical descriptors, and toxicity model of dicamba. ▶ Physicochemical properties are from (*a*) experimental value from Episuite [39], (*b*) estimated values from Episuite [39], (*c*) [44], (*d*) [49]

Because of lack of better information we used Eq. 12 for this case too. However, the validity of the assumptions underlying Eq. 12 for double negatively charged ions is unclear.

In water flea and fish, the metabolite 3,6-dichlorosalicylic acid contributes more to the mixture effect than the parent compound despite its lower fraction, whereas the other two metabolites are insignificant. Note that the uncertainty related to estimating the D_{lipw} (pH7) of the double negatively charged 3,6-dichlorosalicylic acid may bias the overall conclusion.

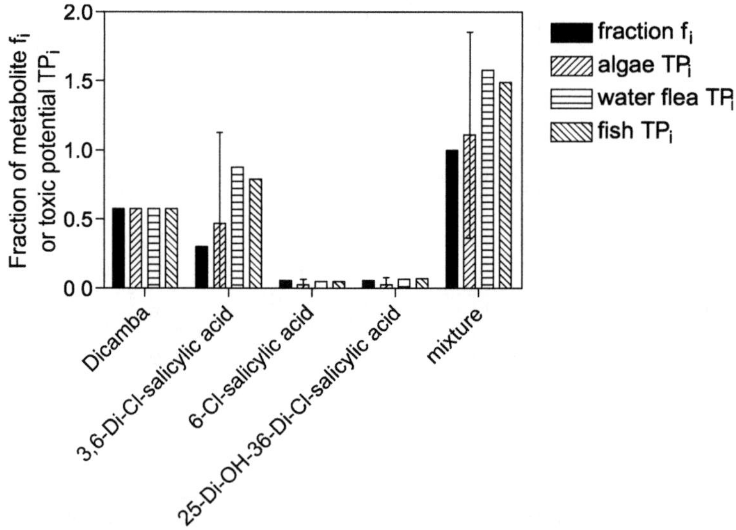

Fig. 9 Fractions of metabolites formed, f_i, and toxic potential of the metabolites, TP_i, and the mixture, TP_{mixture}, for dicamba. *Bars* correspond to the average between $TP_{i,\mathrm{min}}$ and $TP_{i,\mathrm{max}}$ and the *error bars* indicate the range between min and max. If there is no *error bar*, the mode of action is either known or baseline toxicity applies

4.1.5 Bromoxynil

Bromoxynil is a special case because the parent compound bromoxyniloctanoate, although being a baseline toxicant, is highly toxic due to its high hydrophobicity (Fig. 10). The hydrolysis product bromoxynil is the active ingredient and has a specific mode of action: Bromoxynil is a potent inhibitor of Photosystem II and is also an uncoupler of photophosphorylation (i.e. destroys the electrochemical proton gradient formed in the electron transport

Dicamba

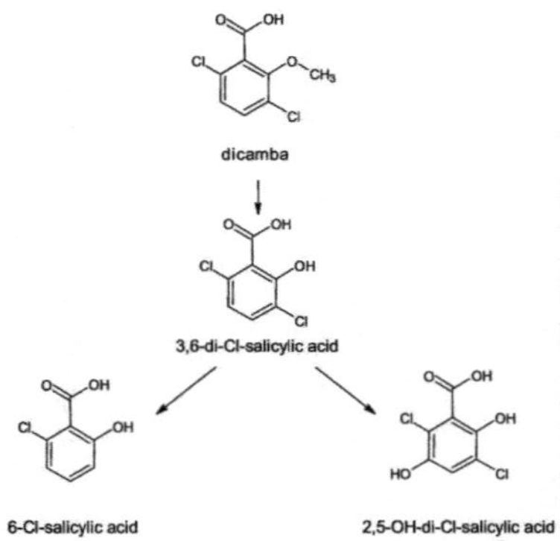

Physicochemical Parameters	pK_a	$\log K_{ow}$		$\log D_{lipw}$(pH 7)
Dicamba	2.55	2.21	(a)	1.51
3,6-Di-Cl-salicylic acid	1.75, 5.36	2.78	(b)	2.03
6-Cl-salicylic acid	2.05, 7.06	2.13	(b)	1.44
2,5-Di-OH-3,6-Di-Cl-salicylic acid	1.70, 5.55	2.30	(b)	1.59

Green algae (*Pseudokirchneriella subcapitata*)	$\log(1/EC_{50}(M))$ baseline	$\log(1/EC_{50}(M))$ experimental		TR		$RP_{i,specific}$ or $RP_{i,max}$	$RP_{i,baseline}$ or $RP_{i,min}$
Dicamba	2.96	6.78	(d)	6539	specific moa	1.00	
3,6-Di-Cl-salicylic acid	3.46	n.a.				3.12	0.0005
6-Cl-salicylic acid	2.90	n.a.				0.86	0.0001
2,5-Di-OH-3,6-Di-Cl-salicylic acid	3.04	n.a.				1.19	0.0002

Water flea (*Daphnia magna*)	$\log(1/EC_{50}(M))$ baseline	$\log(1/EC_{50}(M))$ experimental		TR		$RP_{i,specific}$ or $RP_{i,max}$	$RP_{i,baseline}$ or $RP_{i,min}$
Dicamba	2.97	3.10	(c)	1.4	baseline		1.00
3,6-Di-Cl-salicylic acid	3.44	n.a.					2.94
6-Cl-salicylic acid	2.91	n.a.					0.86
2,5-Di-OH-3,6-Di-Cl-salicylic acid	3.04	n.a.					1.18

Fish (*Pimephales promelas*)	$\log(1/EC_{50}(M))$ baseline	$\log(1/EC_{50}(M))$ experimental		TR		$RP_{i,specific}$ or $RP_{i,max}$	$RP_{i,baseline}$ or $RP_{i,min}$
Dicamba	2.87	n.a.	(c)	n.a.			1.00
3,6-Di-Cl-salicylic acid	3.29	n.a.					2.64
6-Cl-salicylic acid	2.82	n.a.					0.88
2,5-Di-OH-3,6-Di-Cl-salicylic acid	2.94	n.a.					1.16

Fig. 10 Degradation pathway, physicochemical descriptors, and toxicity model of bromoxynil ▶

chain of the photosystem and consequently inhibits ATP production). Thus, the TR of bromoxynil is very high for algae but around 10 for water flea and fish, respectively. Consequently, the moa assignment is ambiguous. Since bromoxynil is the active ingredient, we set it to be the parent compound for the purpose of our calculations (Fig. 10). This is reasonable in the case of algae because in algae bromoxynil also has the highest TR. However, for water flea and fish, this results in very high RP_i values for the considerably more hydrophobic bromoxynil-octanoate.

According to the calculations of Gasser et al. [41] the presence of bromoxynil-octanoate in water is negligible. Also, despite its high specific toxicity bromoxynil does not dominate the toxic potential because its fraction is only 6% (Fig. 11). Instead, the metabolites B-benzamide and B-benzoic acid are the dominant species and also contribute most to the toxic potential of the mixture. However, their mode of action is not defined and therefore the uncertainty in our predictions is large.

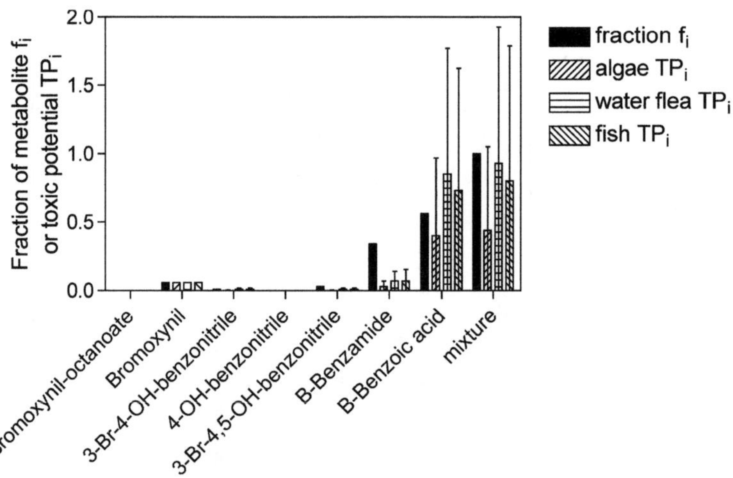

Fig. 11 Fractions of metabolites formed, f_i, and toxic potential of the metabolites, TP_i, and the mixture, $TP_{mixture}$, for bromoxynil. *Bars* correspond to the average between $TP_{i,min}$ and $TP_{i,max}$ and the *error bars* indicate the range between min and max

4.1.6 Chlorothalonil

The fungicide chlorothalonil is used in agriculture and for biocidal treatment of wood. Very little information was available on its transformation products—even the K_{ow} had to be estimated (Fig. 12). However, the analysis

Bromoxynil

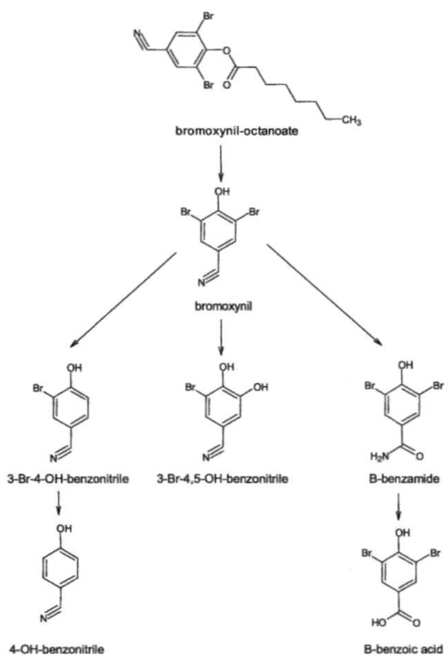

Physicochemical Parameters	pK_a	$\log K_{ow}$		$\log D_{lipw}$(pH 7)
Bromoxynil-octanoate	-	5.9	(c)	5.85
Bromoxynil	4.42	2.97	(d)	2.21
3-Br-4-OH-benzonitrile	6.07	2.50	(b)	2.07
4-OH-benzonitrile	7.77	1.6	(a)	1.90
3-Br-4,5-OH-benzonitrile	5.61, 7.23	2.02	(b)	1.43
B-Benzamide	4.56	2.04	(b)	1.37
B-Benzoic acid	3.52, 5.09	3.17	(b)	2.38

Green algae (*Pseudokirchneriella subcapitata*)	$\log(1/EC_{50}(M))$ baseline	$\log(1/EC_{50}(M))$ experimental		TR		$RP_{i,specific}$ or $RP_{i,max}$	$RP_{i,baseline}$ or $RP_{i,min}$
Bromoxynil-octanoate	7.09	6.16	(e)	0.1	baseline		0.71
Bromoxynil	3.63	6.31	(e,f)	479	specific moa	1.00	
3-Br-4-OH-benzonitrile	3.50			n.a.		0.74	0.0022
4-OH-benzonitrile	3.34			n.a.		0.51	0.0015
3-Br-4,5-OH-benzonitrile	2.89			n.a.		0.18	0.0005
B-Benzamide	2.83			n.a.		0.16	0.0005
B-Benzoic acid	3.79			n.a.		1.45	0.0043

Water flea (*Daphnia magna*)	$\log(1/EC_{50}(M))$ baseline	$\log(1/EC_{50}(M))$ experimental		TR		$RP_{i,specific}$ or $RP_{i,max}$	$RP_{i,baseline}$ or $RP_{i,min}$
Bromoxynil-octanoate	6.88	6.7	(g,h)	1	baseline		484
Bromoxynil	3.32	4.19	(g,h)	7	specific moa	1.00	
3-Br-4-OH-benzonitrile	3.47	n.a.		n.a.		1.42	0.19
4-OH-benzonitrile	3.32	n.a.		n.a.		1.00	0.13
3-Br-4,5-OH-benzonitrile	2.90	n.a.		n.a.		0.38	0.05
B-Benzamide	2.84	n.a.		n.a.		0.33	0.04
B-Benzoic acid	3.75	n.a.		n.a.		2.70	0.36

Fish (*Pimephales promelas*)	$\log(1/EC_{50}(M))$ baseline	$\log(1/EC_{50}(M))$ experimental		TR		$RP_{i,specific}$ or $RP_{i,max}$	$RP_{i,baseline}$ or $RP_{i,min}$
Bromoxynil-octanoate	6.39	n.a.			baseline		112
Bromoxynil	3.19	4.34	(h)		specific moa	1.00	
3-Br-4-OH-benzonitrile	3.33	n.a.				1.37	0.10
4-OH-benzonitrile	3.19	n.a.				1.00	0.07
3-Br-4,5-OH-benzonitrile	2.81	n.a.				0.42	0.03
B-Benzamide	2.76	n.a.				0.37	0.03
B-Benzoic acid	3.58					2.45	0.17

Fig. 12 Degradation pathway, physicochemical descriptors, and toxicity model of chlorothalonil. Physicochemical properties are (*a*) experimental value from Episuite [39], (*b*) estimated values from Episuite [39], (*c*) [50], (*d*) [51], (*e*) [52], (*f*) [49], (*g*) http://europa.eu.int/comm/food/plant/protection/evaluation/existactive/list_bromoxynil, (*h*) [44]

of the parent compound indicated a highly specific effect in all three aquatic species, with a TR as high as 900 for algae. Correspondingly, the uncertainty is rather high for the RP_i estimates of the transformation products; they vary by almost three orders of magnitude between specific and baseline effect. If the transformation products are indeed baseline toxicants, their contribution to the overall toxicity is negligible. If they are however specifically active with the same moa as the parent compound, their contribution to the $TP_{mixture}$ might be relevant.

The parent compound should be readily degradable in the environment [41], thus f_{parent} is 0. The dominant metabolites are 4-OH-2,5,6-trichloroisothalonitrile and 3-cyano-2,4,5,6-tetrachlorobenzamide but due to their low toxicity, the overall $TP_{mixture}$ is small. As the error bars in Fig. 13 show, the uncertainty related to correct classification of moa does not lead

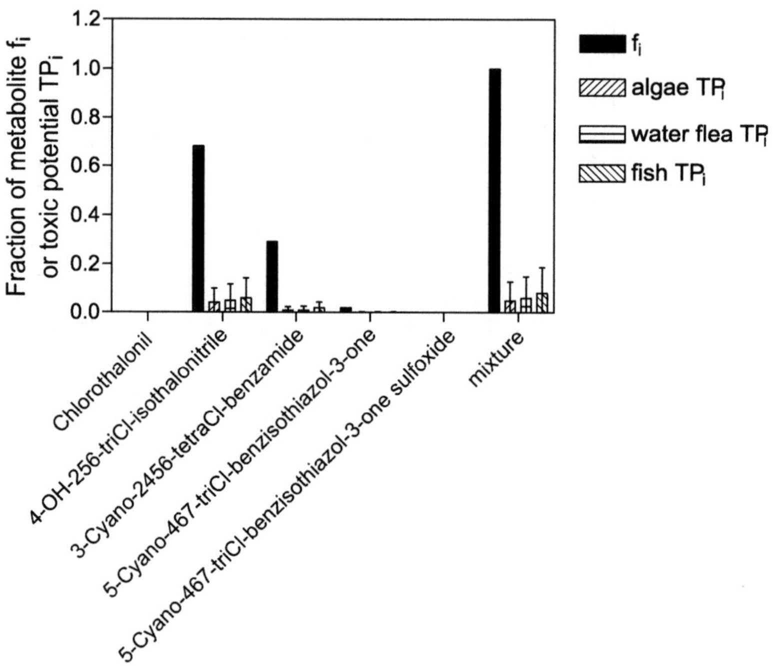

Fig. 13 Fractions of metabolites formed, f_i, and toxic potential of the metabolites, TP_i, and the mixture, $TP_{mixture}$, for chlorothalonil. *Bars* correspond to the average between $TP_{i,min}$ and $TP_{i,max}$ and the *error bars* indicate the range between min and max

Chlorothalonil

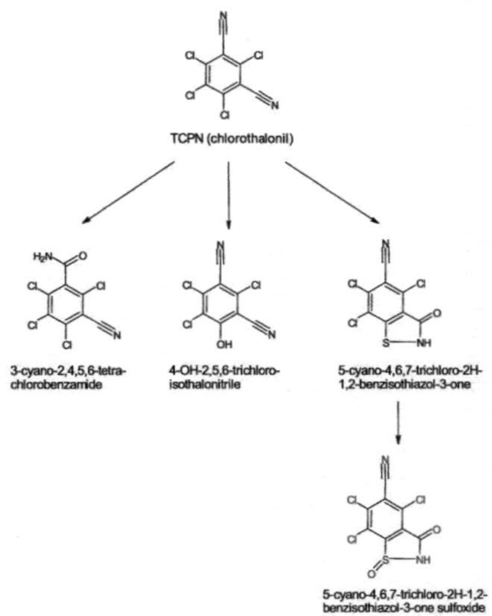

Physicochemical Parameters	pK_a	$\log K_{ow}$		$\log D_{lipw}$(pH 7)
Chlorothalonil	-	3.05	(a)	3.27
4-OH-2,5,6-triCl-isothalonitrile	1.4	3.09	(b)	2.31
3-Cyano-2,4,5,6-tetraCl-benzamide	-	1.73	(b)	2.08
5-Cyano-4,6,7-triCl-benzisothiazol-3-one	-	2.12	(b)	2.43
5-Cyano-4,6,7-triCl-benzisothiazol-3-one sulfoxide	-	1.91	(b)	2.24

Green algae (*Pseudokirchneriella subcapitata*)	$\log(1/EC_{50}$(M)) baseline	$\log(1/EC_{50}$(M)) experimental		TR		$RP_{i,specific}$ or $RP_{i,max}$	$RP_{i,baseline}$ or $RP_{i,min}$
Chlorothalonil	4.64	7.59	(c)	898	specific moa	1.00	
4-OH-2,5,6-triCl-isothalonitrile	3.72	n.a.		n.a.		0.12	0.0001
3-Cyano-2,4,5,6-tetraCl-benzamide	3.51	n.a.		n.a.		0.07	0.0001
5-Cyano-4,6,7-triCl-benzisothiazol-3-one	3.84	n.a.		n.a.		0.16	0.0002
5-Cyano-4,6,7-triCl-benzisothiazol-3-one sulfoxide	3.66	n.a.		n.a.		0.11	0.0001

Water flea (*Daphnia magna*)	$\log(1/EC_{50}$(M)) baseline	$\log(1/EC_{50}$(M)) experimental		TR		$RP_{i,specific}$ or $RP_{i,max}$	$RP_{i,baseline}$ or $RP_{i,min}$
Chlorothalonil	4.55	6.47	(c,d)	83	specific moa	1.00	
4-OH-2,5,6-triCl-isothalonitrile	3.69	n.a.		n.a.		0.14	0.0017
3-Cyano-2,4,5,6-tetraCl-benzamide	3.48	n.a.		n.a.		0.08	0.0010
5-Cyano-4,6,7-triCl-benzisothiazol-3-one	3.80	n.a.		n.a.		0.18	0.0021
5-Cyano-4,6,7-triCl-benzisothiazol-3-one sulfoxide	3.63	n.a.		n.a.		0.12	0.0014

Fish (*Pimephales promelas*)	$\log(1/EC_{50}$(M)) baseline	$\log(1/EC_{50}$(M)) experimental		TR		$RP_{i,specific}$ or $RP_{i,max}$	$RP_{i,baseline}$ or $RP_{i,min}$
Chlorothalonil	4.30	7.07	(e)	591	specific moa	1.00	
4-OH-2,5,6-triCl-isothalonitrile	3.52	n.a.		n.a.		0.17	0.0003
3-Cyano-2,4,5,6-tetraCl-benzamide	3.33	n.a.		n.a.		0.11	0.0002
5-Cyano-4,6,7-triCl-benzisothiazol-3-one	3.62	n.a.		n.a.		0.21	0.0004
5-Cyano-4,6,7-triCl-benzisothiazol-3-one sulfoxide	3.46	n.a.		n.a.		0.15	0.0002

to large absolute errors in the $TP_{mixture}$. This finding contrasts the case of bromoxynil. The difference is due to the very low hydrophobicity of the chlorothalonil transformation products resulting in a small contribution to TP.

4.1.7
Concluding Remarks for the Pesticide Case Studies

Availability of toxicity data for pesticides is good in the case of the herbicides diuron and atrazine and for these two compounds the mixture toxicity predictions appear quite robust. In contrast, for specifically acting pesticides for which the mode of action of the transformation products is not clear, our modeling results in relatively large uncertainty ranges: It covers the entire range from baseline toxicity to specific toxicity. Nevertheless, the model is potentially very useful to decide if transformation products are likely to play a role at all and to focus further ecotoxicity testing on those that might play a role.

4.2
Human Metabolites of Pharmaceuticals

4.2.1
General Remarks

Human and veterinary pharmaceuticals are often highly metabolized before they are excreted into wastewater or the environment. Thus, we have to consider both the metabolism during the pharmacokinetic phase in the target organism and the environmental transformation processes. For simplicity, we focus in the following on the metabolites formed in organisms, but environmental transformation products could be treated in an analogous way as the pesticides discussed above. Also veterinary and human pharmaceuticals can be approached in a similar fashion but the examples below refer to human pharmaceuticals.

The advantage of addressing human metabolites is that there is a vast body of information on the pharmacokinetics available. Therefore, information on metabolic pathways and fractions of metabolites formed can be taken directly from the literature. The disadvantage is that, unlike for pesticides, there are almost no experimental ecotoxicity data available for the metabolites. In addition, the environmental risk assessment of pharmaceuticals should rely exclusively on chronic toxicity data [53] because it is suspected that some pharmaceuticals have unusually high acute-to-chronic ratios, which was confirmed by a recent review [54]. Although the model presented here can in theory easily be used for chronic toxicity data, its practical implementation is limited by the unavailability of QSARs for chronic endpoints. For illustration purposes, we again chose the acute toxicity QSARs presented in Table 2. We have modeled 42 human pharmaceuticals with the proposed method using a different set of QSAR equations [12].

Here, we present a few exemplary pharmaceuticals to illustrate the application of the model to drug metabolites. The first case study of three β-blockers is interesting for a demonstration of how different metabolism patterns influence the expected mixture toxicity and how this is related to absolute effect levels of the different β-blockers. We chose diclofenac and carbamazepine as a second illustrative set because they have been detected in rather high concentrations (in the μg/L range) in wastewater treatment effluents [55]. On the one hand this reflects their abundant usage and on the other hand it reflects the fact that they are typically removed by less than 20% in conventional municipal wastewater treatment plants [56]. Thus, the question arises if the parent compounds are only the tip of the iceberg with their human metabolites contributing substantially to the overall risk. The final example is fluoxetine. This compound was chosen because it is frequently detected in North American surface waters [57] and because its main metabolite, norfluoxetine, is pharmacologically active.

4.2.2
β-Blockers

In an earlier study, we compared the toxic potency of different β-blockers and their human metabolites in the fractions that are excreted in urine and feces [11]. An interesting pattern can be observed. Propranolol is the most hydrophobic β-blocker and accordingly the most toxic one. All β-blockers are slightly more toxic towards algae than towards other aquatic organisms, which points to a specific mode of toxic action in algae and baseline toxicity in water flea and fish [11]. Figure 14A gives an overview of the $TP_{mixture}$ for algae scaled to 100% for each parent compound. It is evident that atenolol, which is only moderately metabolized in the human body, keeps almost all its toxic potential, $TP_{mixture}$, after metabolism. In contrast, propranolol is extensively metabolized in humans and the metabolites were predicted to be less toxic. Therefore, for propranolol the inclusion of metabolism results in a large impact on the environmental risk assessment by strongly reducing the $TP_{mixture}$.

In Fig. 14B, the EC_{50} values of the parent compounds were normalized with the $TP_{mixture}$. For a better appreciation of the effect, the negative logarithm of this ratio is plotted, i.e. a higher number relates to higher toxicity. The parent compound propranolol is much more toxic than metoprolol and atenolol. Consequently, despite the fact that metabolism results in quite a dramatic reduction of the toxic potential, propranolol remains the most ecotoxic of the three β-blockers even if metabolism is included in the toxicity analysis.

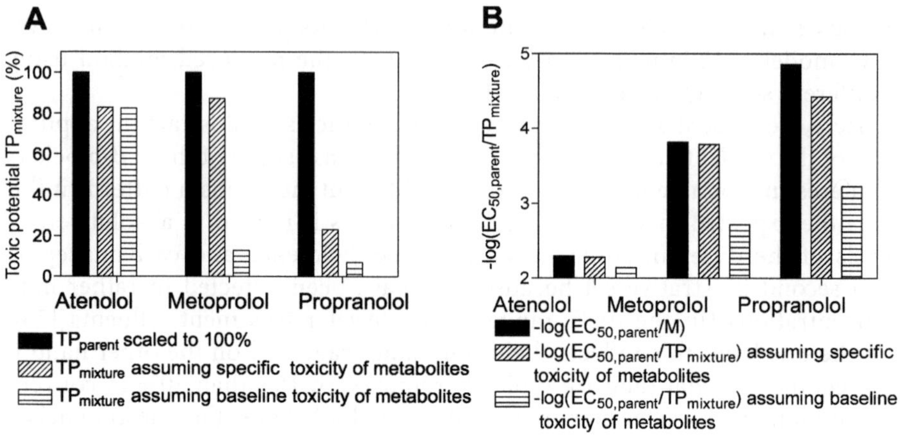

Fig. 14 Toxic potential of three β-blockers towards algae. **A** Modeled mixture effect $TP_{mixture}$ in relation to each parent compound. **B** Toxicity of the mixture calculated by scaling absolute toxicity of the parent compound by the $TP_{mixture}$

4.2.3 Diclofenac

Diclofenac is a nonsteroid anti-inflammatory drug. The main metabolites of diclofenac are the conjugates (here modeled as glucoronides) and 4′-hydroxypropranolol. Other oxidation products occur in traces only [58–60] (Figs. 15 and 16). All these metabolites are about a factor four less toxic than the parent compound, with the exception of the di-hydroxylated diclofenac, whose relative potency was modeled to be only 2% of the parent compound (Fig. 15). Since the parent compound was baseline toxic for all investigated ecotoxicity endpoints (Fig. 15) and for other acute endpoints [17], we also assumed that the metabolites were baseline toxicants.

In accordance with their small relative potencies, the metabolites are predicted to not dominate the $TP_{mixture}$ (Fig. 16) and the $TP_{mixture}$ to be smaller than one, indicating that metabolism in the human body reduces the ecotoxicity of diclofenac.

4.2.4 Carbamazepine

Carbamazepine is an antiepileptic drug, which is relatively persistent in wastewater treatment plants [56]. Therefore, it has a relatively high risk quo-

Fig. 15 Metabolic pathway of diclofenac in humans, physicochemical descriptors, and toxicity model. Physicochemical properties are from (*a*) [61], (*b*) [62], (*c*) [63] ▶

Diclofenac

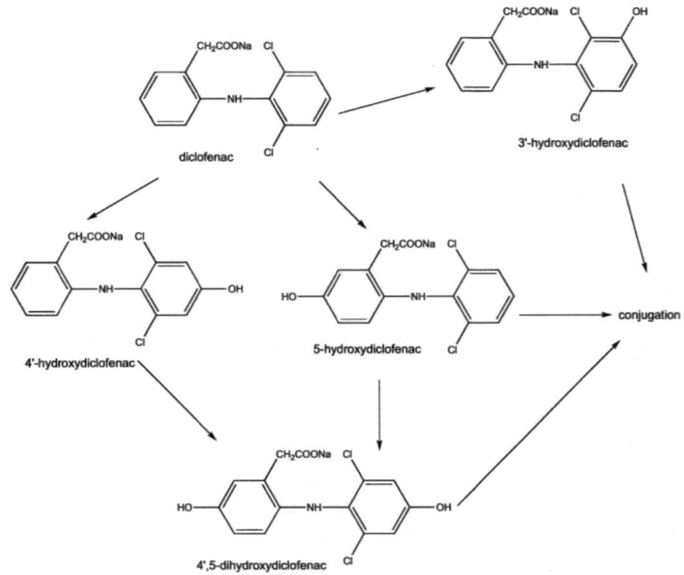

Physicochemical Parameters				
	pK_a	$logK_{ow}$		log D_{lipw}(pH 7)
Diclofenac	3.99	4.51	(a)	3.82
Conjugate (glucoronide)	-	2.68	(b)	3.08
4'-Hydroxydiclofenac	3.99*	3.84	(b)	3.18
3'-Hydroxydiclofenac	3.99*	3.84	(b)	3.18
5-Hydroxydiclofenac	3.99*	3.84	(b)	3.18
4',5-Hydroxydiclofenac	3.99*	2.5	(b)	1.91

* Hydroxy substitution on ring should not have a strong influence on the acidity of the caboxylic acid, and in any case at pH 7 the carboxylic acid is >99.9% in its anionic form, therefore the pK_a of the parent compound was used.

Green algae (*Pseudokirchneriella subcapitata*)	log(1/EC_{50}(M)) baseline	log(1/EC_{50}(M)) experimental		TR	$RP_{i,baseline}$
Diclofenac	5.16	3.64	(c)	0.03	1.00
Conjugate (glucoronide)	4.46	n.a.		baseline	0.20
4'-Hydroxydiclofenac	4.55	n.a.			0.25
3'-Hydroxydiclofenac	4.55	n.a.			0.25
5-Hydroxydiclofenac	4.55	n.a.			0.25
4',5-Hydroxydiclofenac	3.34	n.a.			0.02

Water flea (*Daphnia magna*)	log(1/EC_{50}(M))	log(1/EC_{50}(M))		TR	$RP_{i,baseline}$
Diclofenac	5.05	3.67	(c)	0.04	1.00
Conjugate (glucoronide)	4.38	n.a.		baseline	0.22
4'-Hydroxydiclofenac	4.47	n.a.			0.27
3'-Hydroxydiclofenac	4.47	n.a.			0.27
5-Hydroxydiclofenac	4.47	n.a.			0.27
4',5-Hydroxydiclofenac	3.33	n.a.			0.02

Fish (*Pimephales promelas*)	log(1/EC_{50}(M)) baseline	log(1/EC_{50}(M)) experimental		TR	$RP_{i,baseline}$
Diclofenac	4.74	3.17	(d)	0.03	1.00
Conjugate (glucoronide)	4.14	n.a.		baseline	0.25
4'-Hydroxydiclofenac	4.23	n.a.			0.30
3'-Hydroxydiclofenac	4.23	n.a.			0.30
5-Hydroxydiclofenac	4.23	n.a.			0.30
4',5-Hydroxydiclofenac	3.20	n.a.			0.03

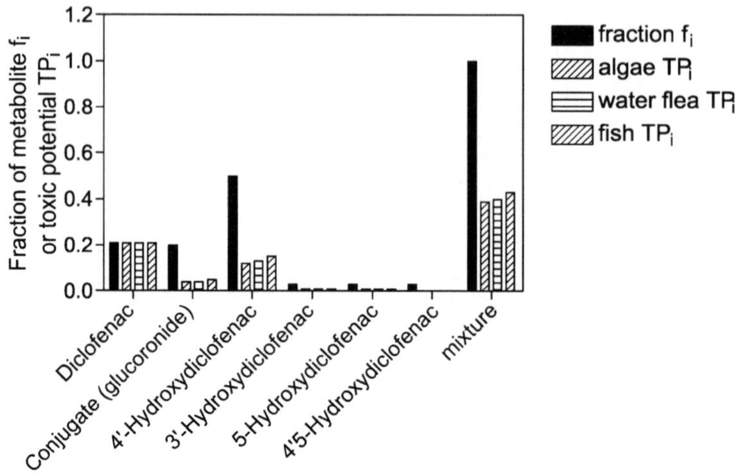

Fig. 16 Fractions of metabolites formed, f_i, and toxic potential of the metabolites, TP_i, and the mixture, $TP_{mixture}$, for diclofenac

Fig. 17 Metabolic pathway of carbamazepine in humans, physicochemical descriptors, and toxicity model. Physicochemical properties are from (*a*) Physprop database www.syrres.com, (*b*) estimated with fragment method according to [11], (*c*) [62], (*d*) [63]

tient for different scenarios [63, 64], despite the fact that it acts only as baseline toxicant in algae, water flea, and fish [62, 63]. It is metabolized by oxidation and conjugation according to the scheme presented in Fig. 17 [59, 60, 65]. The most potent metabolite is iminostilbene, which is predicted to be 20- to 30-times more potent than the parent compound, even if only baseline toxicity of the metabolite is assumed. However, the fraction of excreted iminostilbene, $f_{iminostilbene}$, is maximum only 5% and thus its quantitative contribution is very low (Fig. 17).

Nonetheless, even if only 5% of the ingested carbamazepine were excreted as iminostilbene, it would dominate the $TP_{mixture}$ due to its high potency relative to the parent compound. Thus, carbamazepine is clearly an example of a human pharmaceutical where the metabolite formation and metabolite ecotoxicity should be further investigated and where the environmental persistence and presence of iminostilbene should be explored experimentally.

4.2.5 Fluoxetine

Fluoxetine is an antidepressant and lifestyle drug frequently detected in North American surface waters. Its concentrations usually are only about two orders of magnitude below measurable effects in the aquatic environment [57, 66, 67]. Almost equal amounts of fluoxetine and its pharmacologi-

Carbamazepine

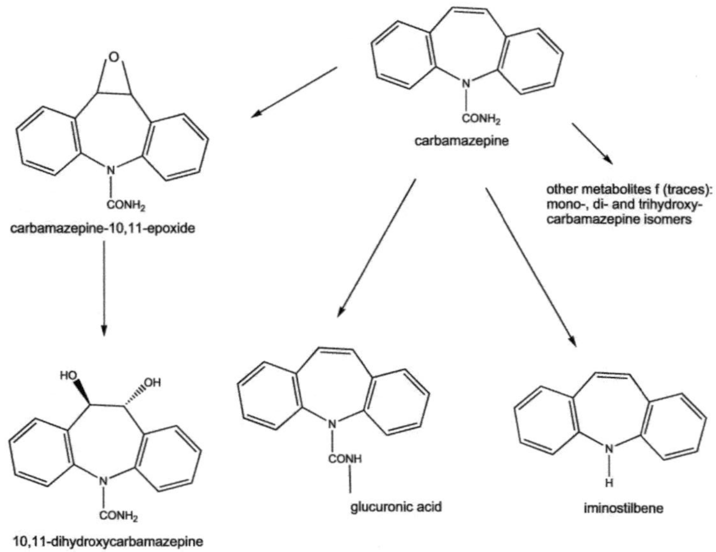

Physicochemical Parameters	pK_a	$logK_{ow}$		log D_{lipw}(pH 7)
Carbamazepine	-	2.45	(a)	2.86
conjugated (glucoronide)	-	1.29	(b)	1.76
Carbamazepine-10, 11-epoxide	-	0.63	(b)	1.13
10,11-Dihydroxycarbamazepine	-	-1.29	(b)	0.47
Iminostilbene	n.a.	4.06	(a)	4.39

Green algae (*Pseudokirchneriella subcapitata*)	$\log(1/EC_{50}(M))$ baseline	$\log(1/EC_{50}(M))$ experimental		TR	$RP_{i,baseline}$
Carbamazepine	4.25	3.50	(c)	0.18	1.00
conjugated (glucoronide)	3.20	n.a.		baseline	0.09
Carbamazepine-10, 11-epoxide	2.60	n.a.			0.02
10,11-Dihydroxycarbamazepine	1.98	n.a.			0.01
Iminostilbene	5.70	n.a.			28.41

Water flea (*Daphnia magna*)	$\log(1/EC_{50}(M))$ baseline	$\log(1/EC_{50}(M))$ experimental		TR	$RP_{i,baseline}$
Carbamazepine	4.18	3.18	(c)	0.10	1.00
conjugated (glucoronide)	3.19	n.a.		baseline	0.10
Carbamazepine-10, 11-epoxide	2.63	n.a.			0.03
10,11-Dihydroxycarbamazepine		n.a.			0.00
Iminostilbene	5.56	n.a.			23.82

Fish (*Pimephales promelas*)	$\log(1/EC_{50}(M))$ baseline	$\log(1/EC_{50}(M))$ experimental		TR	$RP_{i,baseline}$
Carbamazepine	3.97	2.37	(d)	0.03	1.00
conjugated (glucoronide)	3.08	n.a.		baseline	0.13
Carbamazepine-10, 11-epoxide	2.57	n.a.			0.04
10,11-Dihydroxycarbamazepine	2.03	n.a.			0.01
Iminostilbene	5.21	n.a.			17.35

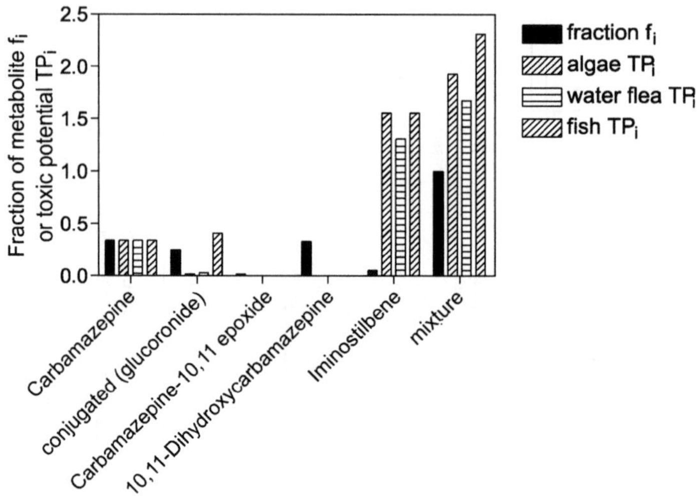

Fig. 18 Fractions of metabolites formed, f_i, and toxic potential of the metabolites, TP_i, and the mixture, $TP_{mixture}$, for carbamazepine

cally active metabolite and demethylation product norfluoxetine are excreted in human urine [68], but hardly anything is known about the environmental fate and effects of norfluoxetine. Both fluoxetine and norfluoxetine inhibit various neuronal ion channels [69] and could therefore also have toxic effects on nontarget aquatic organisms. Another review reported a pharmocokinetic study with radioactively marked fluoxetine [70]. 11% of unchanged fluoxetine and 7% in its conjugated form as well as 7% and 8% of norfluoxetine and its conjugates were excreted (Fig. 19) [70]. Additionally, 20% was converted to hippuric acid [70]. Other authors also reported the presence of trifluoromethylphenol, albeit in a small fraction [59]. Fluoxetine acts according to a specific mode of toxic action in all three aquatic species but the TR is highest for algae with a value of almost 400. This general finding is not unexpected given that fluoxetine is an inhibitor of ion channels. However, this high specific toxicity towards algae is somewhat surprising and warrants further investigation. Since norfluoxetine is even more pharmacologically active than fluoxetine [69], we can safely assume that norfluoxetine also acts specifically. However, we cannot model a potentially higher intrinsic activity. Again, these assumptions in our model have a high inherent uncertainty. Because the abundance of norfluoxetine is about as high as that of fluoxetine, the ecotoxic

Fig. 19 Metabolic pathway of fluoxetine in humans, physicochemical descriptors, and toxicity model. Physicochemical properties are from (*a*) [66] (*b*) Physprop database www.syrres.com, (*c*) estimated with fragment method according to [11], (*d*) calculated using Advanced Chemistry Development (ACD/Labs) Software V8.14 for Solaris ▶

Fluoxetine

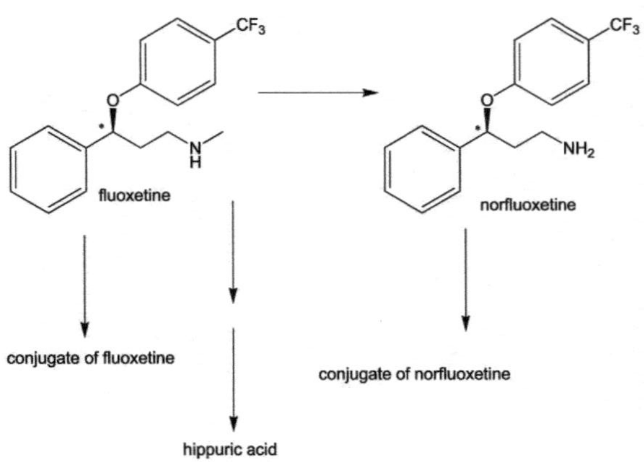

Physicochemical Parameters	pK_a	$logK_{ow}$		log D_{lipw}(pH 7)
Fluoxetine	10.06	4.05	(a,b)	3.18
Conjugated fluoxetine	-	2.89	(c)	3.13
Norfluoxetine	9.05	3.39	(a,b)	2.58
Conjugated norfluoxetine	-	2.23	(c)	2.53
Hippuric acid	3.68	0.31	(b)	-0.20

	log(1/EC_{50}(M)) baseline	log(1/EC_{50}(M)) experimental		TR		$RP_{i,specific}$ or $RP_{i,max}$	$RP_{i,baseline}$ or $RP_{i,min}$
Green algae (*Pseudokirchneriella subcapitata*)							
Fluoxetine	4.55	7.11	(a)	362	specific	1.00	
Conjugated fluoxetine	4.50	n.a.				0.90	0.0025
Norfluoxetine	3.98	n.a.				0.27	0.0007
Conjugated norfluoxetine	3.93	n.a.				0.24	0.0007
Hippuric acid	1.34	n.a.				0.00062	0.000002
Water flea (*Daphnia magna*)							
Fluoxetine	4.47	5.58	(a)	12.82	specific	1.00	
Conjugated fluoxetine	4.43	n.a.				0.90	0.070
Norfluoxetine	3.93	n.a.				0.29	0.022
Conjugated norfluoxetine	3.89	n.a.				0.26	0.020
Hippuric acid	1.43	n.a.				0.0009	0.0001
Fish (*Pimephales promelas*)							
Fluoxetine	4.23	5.64	(a)	25.95	specific	1.00	
Conjugated fluoxetine	4.19	n.a.				0.91	0.035
Norfluoxetine	3.74	n.a.				0.33	0.013
Conjugated norfluoxetine	3.70	n.a.				0.30	0.0115
Hippuric acid	1.49	n.a.				0.0018	0.0001

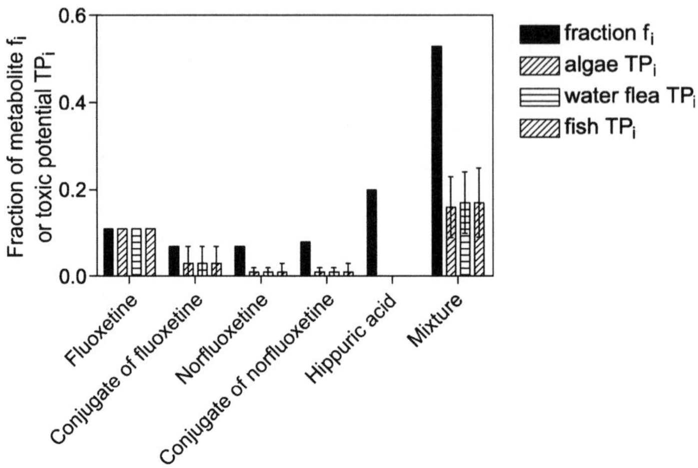

Fig. 20 Fractions of metabolites of fluoxetine formed during human metabolism and excreted via urine and feces (f_i) and toxic potential of the metabolites, TP_i, and the mixture, $TP_{mixture}$. *Bars* correspond to average between $TP_{i,min}$ and $TP_{i,max}$ and the *error bars* indicate the range between min and max

effects of norfluoxetine should be experimentally investigated. Moreover, further environmental degradation of the excreted metabolite cocktail would presumably result in the formation of even more norfluoxetine because deconjugation and demethylation are reactions that are expected to occur also in wastewater treatment plants.

In the simulation of the toxic potential depicted in Fig. 20, we used the excretion data as reported by de Vane [70], whereas in earlier work we had still relied on average data from the different pharmaceutical compendia [12]. There are two options to treat the conjugates. They can be modeled in the form they are excreted. Most conjugates are very hydrophilic and do not contribute much to the TP. Also they are expected to be baseline toxicants. However, some of the conjugates will be cleaved in the wastewater treatment plant, so another option would be to account for them as their unconjugated precursor. Since in this section we are modeling the excreted mixtures only, we account for the metabolites in their excreted form. Under different circumstances one might choose to model the hydrolyzed conjugates. An alternative option would be to set up a two-stage model, where human metabolism is coupled with environmental transformation of all metabolites excreted.

Here we chose the first option and since the excreted metabolites differed significantly between the different literature sources, we chose exclusively the metabolite pattern reported by de Vane [70]. The demethylation of fluoxetine to norfluoxetine has a relatively large impact on the RP_i, which in turn results in a smaller contribution to the overall $TP_{mixture}$. The most abundant metabolite, hippuric acid, does not contribute at all to the $TP_{mixture}$ due to its low relative potency. The conjugates contribute less to the TP than their precursors—even if we assumed all conjugates to be glucoronides, which form the least hydrophilic conjugates. Although the parent compounds dominate the $TP_{mixture}$, the contribution from the metabolites cannot be neglected.

4.2.6 Concluding Remarks for the Pharmaceutical Case Studies

Here, we presented only a few illustrative examples on the assessment of mixtures of pharmaceuticals and their metabolites. A more comprehensive overview on the application of our model to 42 pharmaceuticals [12] shows that, with few exceptions, metabolism in the human body typically decreases the ecotoxic potential of a pharmaceutical. However, this can also be viewed from a different angle: The concentration of the parent compound measured in the environment is only part of the mixture of parent drug and metabolites. Thus, even if the metabolites are presumably less toxic because they are more hydrophilic, they still add to the overall toxic potential of the mixture and must not be a priori neglected. In addition, the fate of more hydrophilic substances can be very different from a more hydrophobic parent so exposure in some compartments may be very different from the parent. The major difficulty in applying the model to pharmaceuticals is that there are virtually no ecotoxicity data available for the metabolites. Therefore, the model only gives very rough estimates.

5 Conclusion and Outlook

We propose a simple and versatile model for the prediction of the toxic potential of mixtures of environmental pollutants and their transformation products. This model assumes concentration addition between the parent compound and its metabolites, which is strictly only applicable if the parent and all metabolites act according to the same mode of toxic action.

The largest uncertainty related to the model is the correct assignment of the appropriate mode of toxic action. Structural alerts, so-called toxicophores, have been suggested to identify compounds with specific moa [30]. However, toxicophores can only provide qualitative information on the presence or absence of a specific moa. Structural alerts can therefore not be used to estimate a metabolite's toxic potential in a quantitative way, but they play an important role in triggering experimental testing.

Although the illustrative cases given here were focused on pesticides and human pharmaceuticals, they are not limited to these compound classes. In earlier work, we have used the same concept for including transformation products into the risk assessment of nonylphenol ethoxylates [13].

As any model, this model is only as good as the input parameters. If there is too little information on the physicochemical properties and the toxicity of the parent and the metabolites, the results will be highly uncertain. Robustness of the model increases with information on the moa of the transformation products.

Despite the limitations, this model may serve for initial hazard estimation and especially for priority setting for experimental investigations. In the case studies we identified a few interesting substances that warrant further exploration. The metabolite norfluoxetine, for instance, should not be neglected in the risk assessment of fluoxetine and should also be monitored in the environment. For the case study pesticides it appears that it is not acceptable to neglect the environmental transformation products. For all pesticides investigated the toxic potential of the mixture of parent compound and its metabolites, $TP_{mixture}$, was higher than that of the parent compound alone, TP_{parent}. This finding is relevant because transformation products are often more commonly found in the environment than the parent compounds [1]. However, extending all monitoring programs to include the relevant transformation products will not always be possible. Thus, modeling the contribution of transformation products, for example with the method proposed here, will be helpful in providing a more realistic risk assessment of environmental pollutants.

Acknowledgements We thank Karin Güdel for collecting pharmacokinetic data. This study was partially funded by the Swiss Federal Office for the Environment (FOEN) in the project KoMet and by the European Union under the 6th framework program in the STREP ERAPharm (SSPI-CT-2003-511135).

References

1. Boxall ABA, Sinclair CJ, Fenner K, Kolpin D, Maud SJ (2004) When synthetic chemicals degrade in the environment. Environ Sci Technol 38:368A–375A
2. Dewez D, Marchand M, Eullaffroy P, Popovic R (2002) Evaluation of the effects of diuron and its derivatives on Lemna gibba using a fluorescence toxicity index. Environ Toxicol 17:493–501
3. Liu WP, Gan JJ, Lee SJ, Werner I (2004) Isomer selectivity in aquatic toxicity and biodegradation of cypermethrin. J Agricultur Food Chem 52:6233–6238
4. Tixier C, Sancelme M, Bonnemoy F, Cuer A, Veschambre H (2001) Degradation products of a phenylurea herbicide, diuron: Synthesis, ecotoxicity, and biotransformation. Environ Toxicol Chem 20:1381–1389
5. Weyandt RG, Pressel J (2003) Strategies for detection of the ecotoxicological risks of degradation metabolites. Anal Bioanal Chem 375:188–189
6. Belfroid AC, van Drunen M, Beek MA, Schrap SM, van Gestel CAM, van Hattum B (1998) Relative risks of transformation products of pesticides for aquatic ecosystems. Sci Total Environ 222:167–183
7. Sinclair CJ, Boxall ABA (2003) Assessing the ecotoxicity of pesticide transformation products. Environ Sci Technol 37:4617–4625

8. Lee JH, Landrum PF (2006) Development of a multi-component damage assessment model (MDAM) for time-dependent mixture toxicity with toxicokinetic interactions. Environ Sci Technol 40:1341–1349
9. Lee JH, Landrum PF (2006) Application of multi-component damage assessment model (MDAM) for the toxicity of metabolized PAH in Hyalella azteca. Environ Sci Technol 40:1350–1357
10. Legierse KCHM, Verhaar HJM, Vaes WHJ, De Bruijn JHM, Hermens JLM (1999) Analysis of the time-dependent acute aquatic toxicity of organophosphorus pesticides: The critical target occupation model. Environ Sci Technol 33:917–925
11. Escher BI, Bramaz N, Richter M, Lienert J (2006) Comparative ecotoxicological hazard assessment of beta-blockers and their human metabolites using a mode-of-action based test battery and a QSAR approach. Environ Sci Technol 40:7402–7408
12. Lienert J, Güdel K, Escher BI (2007) Screening method for ecotoxicological hazard assessment of 42 pharmaceuticals considering human metabolism and excretory routes. Environ Sci Technol 41:4471–4478
13. Fenner K, Kooijman C, Scheringer M, Hungerbuhler K (2002) Including transformation products into the risk assessment for chemicals: The case of nonylphenol ethoxylate usage in Switzerland. Environ Sci Technol 36:1147–1154
14. Villeneuve D, Blankenship A, Giesy J (2000) Derivation and application of relative potency estimates based on in-vitro bioassays. Environ Sci Toxicol 19:2835–2843
15. Altenburger R, Backhaus T, Boedeker W, Faust M, Scholze M, Grimme LH (2000) Predictability of the toxicity of multiple chemical mixtures to Vibrio Fischeri: mixtures composed of similarly acting chemicals. Environ Toxicol Chem 19:2341–2347
16. Backhaus T, Altenburger R, Boedeker W, Faust M, Scholze M, Grimme LH (2000) Predictability of the toxicity of multiple mixtures of dissimilarly acting chemicals to Vibrio Fischeri. Environ Toxicol Chem 19:2348–2356
17. Escher BI, Bramaz N, Eggen RIL, Richter M (2005) In-vitro assessment of modes of toxic action of pharmaceuticals in aquatic life. Environ Sci Technol 39:3090–3100
18. Altenburger R, Walter H, Grote M (2004) What contributes to the combined effect of a complex mixture? Environ Sci Technol 38:6353–6362
19. Junghans M, Backhaus T, Faust M, Scholze M, Grimme LH (2006) Application and validation of approaches for the predictive hazard assessment of realistic pesticide mixtures. Aqua Toxicol 76:93–110
20. McCarty LS, Hodson PV, Craig GR, Kaiser KLE (1985) The use of quantitative structure-activity relationships to predict the acute and chronic toxicities of organic chemicals to fish. Environ Tox Chem 4:595–606
21. Hermens JLM (1989) Quantitative structure-activity relationships of environmental pollutants. In The Handbook of Environmental Chemistry, Reaction and Processes; Hutzinger J (ed) Springer, Berlin, Germany, vol 2E, pp 111–162
22. Hansch C, Leo A (1995) Exploring QSAR Fundamentals and Applications in Chemistry and Biology. Am Chem Soc, Washington, DC
23. van Wezel AP, Opperhuizen A (1995) Narcosis due to environmental pollutants in aquatic organisms: residue-based toxicity, mechanisms, and membrane burdens. Crit Rev Toxicol 25:255–279
24. Lipnick RL, Watson K, Strausz A (1987) A QSAR study of the acute toxicity of some industrial organic chemicals to goldfish. Narcosis, electrophile and proelectrophile mechanism. Xenobiotica 17:1011–1025
25. Verhaar HJM, Ramos EU, Hermens JLM (1996) Classifying environmental pollutants. 2: Separation of class 1 (baseline toxicity) and class 2 (polar narcosis) type compounds based on chemical descriptors. J Chemometrics 10:149–162

26. Maeder V, Escher BI, Scheringer M, Hungerbühler K (2004) Toxic ratio as an indicator of the intrinsic toxicity in the assessment of persistent, bioaccumulative, and toxic chemicals. Environ Sci Technol 38:3659–3666
27. Bradbury SP (1994) Predicting modes of toxic action from chemical structure. An overview. SAR and QSAR in Environ Res 2:89–104
28. Hermens JLM, Balaz S, Damborsky J, Karcher W, Müller M, Sabljic A, Sjöström M (1995) Assessment of QSARs for predicting fate and effects of chemicals in the environment: an international European project. SAR and QSAR Environ Res 3:223–236
29. Spycher S, Nendza M, Gasteiger J (2004) Comparison of different classification methods applied to a mode of toxic action data set. QSAR Comb Sci 23:779–791
30. von der Ohe PC, Kühne R, Ebert R-U, Altenburger R, Liess M, Schüürmann G (2005) Structural alerts- a new classification model to discriminate excess toxicity from narcotic effect levels of organic compounds in the acute daphnid assay. Chem Res Toxicol 18:536–555
31. European Commission Technical Guidance Document in Support of Commission Directive 93/67/EEC on Risk Assessment for New Notified Substances, Commission Regulation (EC) No 1488/94 on Risk Assessment for Existing Substances, and Directive 98/8/EC of the European Parliament and of the Council Concerning the Placing of Biocidal Products on the Market, Office for Official Publications of the European Communities, 2003. http://ecb.jrc.it/
32. Vaes WHM, Urrestarazu-Ramos E, Verhaar H, Hermens JLM (1998) Acute toxicity of nonpolar versus polar narcosis: is there a difference? Environ Toxicol Chem 17:1380–1384
33. Vaes WHJ, Urrestarazu-Ramos E, Hamwick C, van Holstein I, Blaauboer BJ, Seinen W, Verhaar HJM, Hermens JLM (1997) Solid phase microextraction as a tool to determine membrane/water partition coefficients and bioavailable concentrations in in-vitro systems. Chem Res Toxicol 10:1067–1072
34. Vaes WHJ, Urrestarazu-Ramos E, Verhaar HJM, Cramer CJ, Hermens JLM (1998) Understanding and estimating membrane/water partition coefficients: Approaches to derive quantitative structure property relationships. Chem Res Toxicol 11:847–854
35. Escher BI, Sigg L (2004) Chemical Speciation of Organics and of Metals at Biological Interfaces. In: Van Leeuwen HP, Köster W (eds) Physicochemical Kinetics and Transport at Biointerfaces. Wiley, Chichester, vol 9, pp 205–271
36. Verhaar HJM, Mulder W, Hermens JLM (1995) QSARs for ecotoxicity, Report prepared within the framework of the project QSAR for Prediction of Fate and Effects of Chemicals in the Environment, an international project of the Environmental Technologies RTD Programme (DGXII/D-1) of the European Commission under Contract number EV5V-CT92_0211
37. Van Leeuwen CJ, van der Zandt PTJ, Aldenberg T, Verhaar HJM, Hermens JLM (1992) Application of QSARs, extrapolation and equilibrium partitioning in aquatic effect assessment. 1. Narcotic industrial pollutants. Environmental Toxicology and Chemistry 11:267–282
38. Hilal SH, Karickhoff SW, Carreira LA (2005) University of Georgia, Athens, GA
39. USEPA Exposure Assessment Tools and Models, United States Environmental Protection Agency, 2005. http://www.epa.gov/oppt/exposure/docs/episuite.htm
40. Hansch C, Leo A, Hoekman D (1995) Exploring QSAR Hydrophobic, Electronic and Steric Constants; American Chemical Society: Washington, DC
41. Gasser L, Fenner K, Scheringer M (2007) Indicators for the exposure assessment of transformation products of organic micropollutants. Environ Sci Technol 41:2445–2451

42. Escher B, Schwarzenbach RP (2002) Mechanistic studies on baseline toxicity and uncoupling as a basis for modeling internal lethal concentrations in aquatic organisms. Aquat Sci 64:20–35
43. USEPA Reregistration Eligibility Decision for Diuron List A Case 0046, United States Environmental Protection Agency, Office of Prevention, Pesticides and Toxic Substances, 2003. http://www.epa.gov/REDs/diuron_red.pdf
44. USEPA ECOTOX Database, United States Environmental Protection Agency, 2006. http://www.epa.gov/ecotox/ecotox_home.htm
45. Fernández-Alba AR, Hernando MD, Piedra L, Chisti Y (2002) Toxicity evaluation of single and mixed antifouling biocides measured with acute toxicity bioassays. Analytica Chimica Acta 456:303–312
46. Urrestarazu-Ramos E, Vaal MA, Hermens JLM (2002) Interspecies sensitivity in the aquatic toxicity of aromatic amines. Environ Toxicol Pharmacol 11:149–158
47. Solomon KR, Baker DB, Richards RP, Dixon DR, Klaine SJ, LaPoint TW, Kendall RJ, Weisskopf CP, Giddings JM, Giesy JP, Hall LW, Williams WM (1996) Ecological risk assessment of atrazine in North American surface waters. Environ Toxicol Chem 15:31–74
48. Stratton GW (1984) Effects of the Herbicide Atrazine and its Degradation Products, Alone and in Combination, on Phototrophic Microorganisms. Arch Environ Contam Toxicol 13:35–42
49. Fairchild JF, Ruessler DS, Haverland PS, Carlson AR (1997) Comparative Sensitivity of Selenastrum capricornutum and Lemna minor to Sixteen Herbicides. Arch Environ Contam Toxicol 32:353–357
50. EuropeanCommission Bromoxynil SANCO/4347/2000 – final, 2004. http://europa.eu.int/comm/food/plant/protection/evaluation/existactive/list_bromoxynil.pdf
51. Escher BI, Hunziker RW, Schwarzenbach RP (2001) Interaction of phenolic uncouplers in binary mixtures: concentration-additive and synergistic effects. Environ Sci Technol 35:3905–3914
52. Peterson HG, Boutin C, Martin PA, Freemark KE, Ruecker NJ, Moody MJ (1994) Aquatic phyto-toxicity of 23 pesticides applied at Expected Environmental Concentrations. Aquatic Toxicol 28:275–292
53. EMEA Guideline on the Environmental Risk Assessment of Medicinal Products for Human Use CHMP/SWP/4447/00, The European Agency for the Evaluation of Medicinal Products, 2006. http://www.emea.eu.int/pdfs/human/swp/444700en.pdf
54. Crane M, Watts C, Boucard T (2006) Chronic aquatic environmental risks from exposure to human pharmaceuticals. Sci Tot Environ 367:23–41
55. Huschek G, Hansen PD (2006) Ecotoxicological classification of the Berlin river system using bioassays in respect to the European Water Framework Directive. Environmental Monitoring and Assessment 121:15–31
56. Joss A, Zabczynski S, Göbel A, Hoffmann B, Löffler D, McArdell CS, Ternes TA, Thomson A, Siegrist H (2006) Biological degradation of pharmaceuticals in municipal wastewater treatment: proposing a classification scheme. Water Research 40:1686–1696
57. Kolpin DW, Furlong ET, Meyer MT, Thurman EM, Zaugg SD (2002) Pharmaceuticals, hormones and other organic contaminants in U.S. streams, 1999–2000: a national reconnaissance. Environ Sci Technol 36:1201–1211
58. Davies MN (1998) Clinical pharmacokinetics of ibuprofen. Clinical Pharmacokinetics 34:101–154

59. Dollery C, Boobis AR, Burley D, Davies DM, Davies DS, Harrison PI, Orme MLE, Park BK, Goldberg LI (eds) (1991) Therapeutic drugs; Churchill Livingstone, New York, NY
60. Baselt RC, Cravey RH (2000) Disposition of Toxic Drugs and Chemicals in Man; fourth edition ed, Chemical Toxicology Institute: Foster City, CA, USA
61. Avdeef A, Box KJ, Comer JEA, Hibbert C, Yam KY (1998) PH-metric log *P*10. Determination of liposomal membrane-water partition coefficients of ionizable drugs. Pharm Res 15:209–215
62. Cleuvers M (2003) Aquatic ecotoxicity of pharmaceuticals including the assessment of combination effects. Toxicol Lett 142:185–194
63. Hanisch B, Abbas B, Kratz W (2002) Ökotoxikologische Bewertung von Humanarzneimitteln in aquatischen Ökosystemen, Landesumweltamt Brandenburg
64. Ferrari B, Mons R, Vollat B, Fraysse B, Paxeus N, Lo Giudice R, Pollio A, Garric J (2004) Environmental risk assessment of six human pharmaceuticals: Are the current environmental risk assessment procedures sufficient for the protection of the aquatic environment? Environ Toxicol Chem 23:1344–1354
65. Saller R (1983) Praktische Pharmakologie. Eigenschaften gebräuchlicher Medikamente; Schattauer: Stuttgart, Germany
66. Brooks BW, Foran CM, Richards SM, Weston J, Turner PK, Stanley JK, Solomon KR, Slattery JK, LaPoint TW (2003) Aquatic Ecotoxicology of fluoxetine. Toxicol Lett 142:169–183
67. Christensen AM, Faaborg-Andersen S, Ingerslev F, Baun A (2007) Mixture and single-substance toxicity of selective serotonin reuptake inhibitors toward algae and crustaceans. Environ Toxicol Chem 26:85–91
68. Berzas Nevado JJ, Villasenor Llenera MJ, Guiberteau Cabanillas C, Rodriguez Robledo V (2006) Screening of citalopram, fluoxetine and their metabolites in human urine samples by gas-chromatography-mass spectrometry. A global robustness/ruggedness study. J Chromatogr A 1123:130–133
69. Traboulsie A, Chemin J, Kupfer E, Nargeot J, Lory P (2006) T-Type calcium channels are inhibited by fluoxetine and its metabolite norfluoxetine. Mol Pharmacol 69:1963–1968
70. de Vane CL (1999) Metabolism and pharmacokinetics of selective serotonin reuptake inhibitors. Cell Mol Neurobiol 19:443–466

Subject Index

MIX
Papier aus verantwortungsvollen Quellen
Paper from responsible sources
FSC® C105338

If you have any concerns about our products,
you can contact us on
ProductSafety@springernature.com

In case Publisher is established outside the EU,
the EU authorized representative is:
Springer Nature Customer Service Center GmbH
Europaplatz 3, 69115 Heidelberg, Germany

Printed by Libri Plureos GmbH
in Hamburg, Germany